Reviews of Environmental Contamination and Toxicology

VOLUME 201

Reviews of Environmental Contamination and Toxicology

Editor
David M. Whitacre

Editorial Board
Lilia A. Albert, Xalapa, Veracruz, Mexico • Charles P. Gerba, Tucson, Arizona, USA
John Giesy, Saskatoon, Saskatchewan, Canada • O. Hutzinger, Bayreuth, Germany
James B. Knaak, Getzville, New York, USA
James T. Stevens, Winston-Salem, North Carolina, USA
Ronald S. Tjeerdema, Davis, California, USA • Pim de Voogt, Amsterdam, The Netherlands
George W. Ware, Tucson, Arizona, USA

Founding Editor
Francis A. Gunther

VOLUME 201

Coordinating Board of Editors

DR. DAVID M. WHITACRE, *Editor*
Reviews of Environmental Contamination and Toxicology

5115 Bunch Road
Summerfield North, Carolina 27358, USA
(336) 634-2131 (PHONE and FAX)
E-mail: dmwhitacre@triad.rr.com

DR. HERBERT N. NIGG, *Editor*
Bulletin of Environmental Contamination and Toxicology

University of Florida
700 Experiment Station Road
Lake Alfred, Florida 33850, USA
(863) 956-1151; FAX (941) 956-4631
E-mail: hnn@LAL.UFL.edu

DR. DANIEL R. DOERGE, *Editor*
Archives of Environmental Contamination and Toxicology

7719 12th Street
Paron, Arkansas 72122, USA
(501) 821-1147; FAX (501) 821-1146
E-mail: AECT_editor@earthlink.net

ISBN: 978-1-4419-0031-9   e-ISBN: 978-1-4419-0032-6
DOI: 10.1007/978-1-4419-0032-6
Springer Dordrecht Heidelberg London New York

Library of Congress Control Number: 2008929339

© Springer Science+Business Media, LLC 2009
All rights reserved. This work may not be translated or copied in whole or in part without the written permission of the publisher (Springer Science+Business Media, LLC, 233 Spring Street, New York, NY 10013, USA), except for brief excerpts in connection with reviews or scholarly analysis. Use in connection with any form of information storage and retrieval, electronic adaptation, computer software, or by similar or dissimilar methodology now known or hereafter developed is forbidden.
The use in this publication of trade names, trademarks, service marks, and similar terms, even if they are not identified as such, is not to be taken as an expression of opinion as to whether or not they are subject to proprietary rights.
While the advice and information in this book are believed to be true and accurate at the date of going to press, neither the authors nor the editors nor the publisher can accept any legal responsibility for any errors or omissions that may be made. The publisher makes no warranty, express or implied, with respect to the material contained herein.

Printed on acid-free paper

Springer is part of Springer Science+Business Media (www.springer.com)

Foreword

International concern in scientific, industrial, and governmental communities over traces of xenobiotics in foods and in both abiotic and biotic environments has justified the present triumvirate of specialized publications in this field: comprehensive reviews, rapidly published research papers and progress reports, and archival documentations. These three international publications are integrated and scheduled to provide the coherency essential for nonduplicative and current progress in a field as dynamic and complex as environmental contamination and toxicology. This series is reserved exclusively for the diversified literature on "toxic" chemicals in our food, our feeds, our homes, recreational and working surroundings, our domestic animals, our wildlife and ourselves. Tremendous efforts worldwide have been mobilized to evaluate the nature, presence, magnitude, fate, and toxicology of the chemicals loosed upon the earth. Among the sequelae of this broad new emphasis is an undeniable need for an articulated set of authoritative publications, where one can find the latest important world literature produced by these emerging areas of science together with documentation of pertinent ancillary legislation.

Research directors and legislative or administrative advisers do not have the time to scan the escalating number of technical publications that may contain articles important to current responsibility. Rather, these individuals need the background provided by detailed reviews and the assurance that the latest information is made available to them, all with minimal literature searching. Similarly, the scientist assigned or attracted to a new problem is required to glean all literature pertinent to the task, to publish new developments or important new experimental details quickly, to inform others of findings that might alter their own efforts, and eventually to publish all his/her supporting data and conclusions for archival purposes.

In the fields of environmental contamination and toxicology, the sum of these concerns and responsibilities is decisively addressed by the uniform, encompassing, and timely publication format of the Springer triumvirate:

Reviews of Environmental Contamination and Toxicology [Vol. 1 through 97 (1962–1986) as Residue Reviews] for detailed review articles concerned with any aspects of chemical contaminants, including pesticides, in the total environment with toxicological considerations and consequences.

Bulletin of Environmental Contamination and Toxicology (Vol. 1 in 1966) for rapid publication of short reports of significant advances and discoveries in the fields of air, soil, water, and food contamination and pollution as well as methodology and other disciplines concerned with the introduction, presence, and effects of toxicants in the total environment.

Archives of Environmental Contamination and Toxicology (Vol. 1 in 1973) for important complete articles emphasizing and describing original experimental or theoretical research work pertaining to the scientific aspects of chemical contaminants in the environment.

Manuscripts for Reviews and the Archives are in identical formats and are peer reviewed by scientists in the field for adequacy and value; manuscripts for the *Bulletin* are also reviewed, but are published by photo-offset from camera-ready copy to provide the latest results with minimum delay. The individual editors of these three publications comprise the joint Coordinating Board of Editors with referral within the Board of manuscripts submitted to one publication but deemed by major emphasis or length more suitable for one of the others.

Coordinating Board of Editors

Preface

The role of Reviews is to publish detailed scientific review articles on all aspects of environmental contamination and associated toxicological consequences. Such articles facilitate the often-complex task of accessing and interpreting cogent scientific data within the confines of one or more closely related research fields.

In the nearly 50 years since *Reviews of Environmental Contamination and Toxicology* (formerly *Residue Reviews*) was first published, the number, scope and complexity of environmental pollution incidents have grown unabated. During this entire period, the emphasis has been on publishing articles that address the presence and toxicity of environmental contaminants. New research is published each year on a myriad of environmental pollution issues facing peoples worldwide. This fact, and the routine discovery and reporting of new environmental contamination cases, creates an increasingly important function for *Reviews*.

The staggering volume of scientific literature demands remedy by which data can be synthesized and made available to readers in an abridged form. Reviews addresses this need and provides detailed reviews worldwide to key scientists and science or policy administrators, whether employed by government, universities or the private sector.

There is a panoply of environmental issues and concerns on which many scientists have focused their research in past years. The scope of this list is quite broad, encompassing environmental events globally that affect marine and terrestrial ecosystems; biotic and abiotic environments; impacts on plants, humans and wildlife; and pollutants, both chemical and radioactive; as well as the ravages of environmental disease in virtually all environmental media (soil, water, air). New or enhanced safety and environmental concerns have emerged in the last decade to be added to incidents covered by the media, studied by scientists, and addressed by governmental and private institutions. Among these are events so striking that they are creating a paradigm shift. Two in particular are at the center of ever-increasing media as well as scientific attention: bioterrorism and global warming. Unfortunately, these very worrisome issues are now super-imposed on the already extensive list of ongoing environmental challenges.

The ultimate role of publishing scientific research is to enhance understanding of the environment in ways that allow the public to be better informed. The term "informed public" as used by Thomas Jefferson in the age of enlightenment conveyed the thought of soundness and good judgment. In the modern sense, being "well informed" has the narrower meaning of having access to sufficient information. Because the public still gets most of its information on science and technology from TV news and reports, the role for scientists as interpreters and brokers of scientific information to the public will grow rather than diminish. Environmentalism is the newest global political force, resulting in the emergence of multi-national consortia to control pollution and the evolution of the environmental ethic. Will the new politics of the 21st century involve a consortium of technologists and environmentalists, or a progressive confrontation? These matters are of genuine concern to governmental agencies and legislative bodies around the world.

For those who make the decisions about how our planet is managed, there is an ongoing need for continual surveillance and intelligent controls, to avoid endangering the environment, public health, and wildlife. Ensuring safety-in-use of the many chemicals involved in our highly industrialized culture is a dynamic challenge, for the old, established materials are continually being displaced by newly developed molecules more acceptable to federal and state regulatory agencies, public health officials, and environmentalists.

Reviews publishes synoptic articles designed to treat the presence, fate, and, if possible, the safety of xenobiotics in any segment of the environment. These reviews can either be general or specific, but properly lie in the domains of analytical chemistry and its methodology, biochemistry, human and animal medicine, legislation, pharmacology, physiology, toxicology and regulation. Certain affairs in food technology concerned specifically with pesticide and other food-additive problems may also be appropriate.

Because manuscripts are published in the order in which they are received in final form, it may seem that some important aspects have been neglected at times. However, these apparent omissions are recognized, and pertinent manuscripts are likely in preparation or planned. The field is so very large and the interests in it are so varied that the Editor and the Editorial Board earnestly solicit authors and suggestions of underrepresented topics to make this international book series yet more useful and worthwhile.

Justification for the preparation of any review for this book series is that it deals with some aspect of the many real problems arising from the presence of foreign chemicals in our surroundings. Thus, manuscripts may encompass case studies from any country. Food additives, including pesticides, or their metabolites that may persist into human food and animal feeds are within this scope. Additionally, chemical contamination in any manner of air, water, soil, or plant or animal life is within these objectives and their purview.

Manuscripts are often contributed by invitation. However, nominations for new topics or topics in areas that are rapidly advancing are welcome. Preliminary communication with the Editor is recommended before volunteered review manuscripts are submitted.

Summerfield, North Carolina   D.M.W.

Contents

1 **Monitoring and Reducing Exposure of Infants to Pollutants in House Dust** 1
John W. Roberts, Lance A. Wallace, David E. Camann, Philip Dickey, Steven G. Gilbert, Robert G. Lewis, and Tim K. Takaro

2 **Pulmonary Toxicity and Environmental Contamination: Radicals, Electron Transfer, and Protection by Antioxidants** 41
Peter Kovacic and Ratnasamy Somanathan

3 **Risk Assessment of *Pseudomonas aeruginosa* in Water** 71
Kristina D. Mena and Charles P. Gerba

4 **Non-thermal Plasmas Chemistry as a Tool for Environmental Pollutants Abatement** 117
Yan-hong Bai, Jie-rong Chen, Xiao-yong Li, and Chun-hong Zhang

5 **Environmental Fate and Global Distribution of Polychlorinated Biphenyls** 137
Angelika Beyer and Marek Biziuk

Index 159

Monitoring and Reducing Exposure of Infants to Pollutants in House Dust

John W. Roberts, Lance A. Wallace, David E. Camann, Philip Dickey, Steven G. Gilbert, Robert G. Lewis, and Tim K. Takaro

Contents

1	Introduction	1
2	Monitoring Pollutants in Surface and Deep Dust	3
	2.1 Monitoring Methods	4
	2.2 Metals	5
	2.3 Pesticides	6
	2.4 PAHs and PCBs	8
	2.5 Phthalates	13
	2.6 PBDEs	14
	2.7 Phenols and Alkylphenols	17
	2.8 Dust Mites, Mold, Other Allergens, Viruses, and Bacteria	18
	2.9 Hygiene Hypothesis	18
3	Cleaning Practices, Carpets, and Safer Cleaning Products	19
	3.1 Cleaning	19
	3.2 Carpets and Alternatives	20
	3.3 Vacuum Cleaners	20
	3.4 Hand Washing	23
	3.5 Safer Cleaning Products	23
4	Reducing Exposure and Health Costs with Home Visits	24
	4.1 Home Surveys	24
	4.2 Reducing Asthma and Health Costs	25
5	Discussion	28
6	Research Recommendations	29
7	Summary	29
	References	30

1 Introduction

Babies come with great potential but great vulnerability. It is estimated that infants eat twice as much dust (100 mg vs. 50 mg/d), weigh one sixth as much, and are up to ten times more vulnerable than are adults to dust exposure (U.S. EPA 2002,

---

L.A. Wallace (✉)
U.S. Environmental Protection Agency, retired, 11568 Woodhollow Court,
Reston, VA, 22091, USA
e-mail: lwallace73@verizon.net

2003). The developing neurological, immune, digestive, and other bodily systems of infants are easily affected at low doses and these systems are less able to metabolize, detoxify, and excrete pollutants (Grandjean and Landrigan 2006; U.S. EPA 1996, 2002, 2003). Up to 11% of toddlers may exhibit pica behavior, eating nonfood items, and may consume up to 10 g of soil and dust per day (Calabrese and Stanek 1991; Mahaffey and Annest 1985). The time of life when exposure occurs may be as important as the dose (Grandjean and Landrigan 2006; Louis et al. 2007). Infants breathe more air, drink and eat more relative to their body weight, and engage in risky behaviors such as mouthing hands, toys, furniture, and other nonfood items. They crawl on floors, where they are in close proximity to carpets, and may breathe higher levels of dust (Rodes et al. 1996). Exposures early in life may trigger sensitization leading to development of chronic diseases such as asthma or predispose to cancer that takes decades to develop (Louis et al. 2007).

Childhood chronic health conditions that limited daily activity by at least 3 mon each year increased from 1.8% in 1960 to 7% in 2004. Changes in physical and social environmental exposures may be a significant cause of this rapid rise (Perrin et al. 2007). The leading chronic diseases of children are asthma, which has increased sharply in the last 30 yr; attention deficit hyperactivity disorder (ADHD), which now affects around 6% of school age children; and obesity, which increased from 5% in 1971–1974 to 18% of children and adolescents in 2002 (Perrin et al. 2007). Not all children with these diseases have physical activity limitations lasting at least 3 mon each year, and therefore are not included in the 7% of children with such conditions. This increase in children's chronic disease has sobering implications for future health costs, school achievement, and work productivity. Krieger et al. (2005) and Takaro et al. (2004) suggest that morbidity from moderate and severe (poorly controlled) asthma can be reduced by 50% or more by reducing exposure to triggers in the home. Braun et al. (2006) suggest that reducing the lead exposure of babies may reduce ADHD. Exposure to neurotoxic chemicals is associated with ADHD, neurodevelopment disorders, autism, loss of intelligence, and mental retardation (Grandjean and Landrigan 2006).

Research suggests that house dust is the main source of infant exposure to allergens (Pope et al. 1993), lead (Davies et al. 1990; Lanphear 1996; Lanphear et al. 2005; U.S. EPA 1997) and polybrominated diphenyl ethers (PBDEs) (Jones-Otazo et al. 2005; Stapleton et al. 2005; Wu et al. 2007). Dust is also a major in-home exposure source for pesticides, polyaromatic hydrocarbons (PAHs), phthalates, endocrine disrupting compounds (EDCs), arsenic, cadmium, chromium, mold, endotoxin, and bacteria (Benson 1985; Butte and Heinzow 2002; Camann et al. 2002a; Lioy 2006; Rasmussen et al. 2001; Roberts et al. 1999; Rudel et al. 2003). The track-in or inside generation of animal feces, hair, and saliva provides a source of viruses, gram-negative bacteria, and endotoxin in dust (Benson 1985; McCaustland et al. 1982). Franke et al. (1997) and Roberts and Ott (2006) suggest that floor dust is a source of indoor air exposure to particles, gram-negative bacteria, volatile organic compounds, and mold. Over 100 potentially toxic metals, pesticides, other carcinogens, other neurotoxins, allergens, and EDCs have been identified in house dust (Butte and Heinzow 2002; Lioy 2006; Papadopoulos 1998; Rudel et al. 2003).

House dust contains many mutagens including direct mutagens that do not require metabolic activation (Roberts et al. 1987; Maertens et al. 2004). Maertens et al. (2008a, b) estimate that 25% of the mutagenic activity in house dust comes from PAHs. The combined risk from this mixture of pollutants in dust is unknown (Menzie et al. 2007).

The fine particles produced by combustion tend to combine with the larger particles found in soil and house dust (Lewis et al. 1999). Soil particles greater than $2\,\mu m$ in diameter are preferentially tracked into homes. Contaminants adhering to such particles include metals, pesticides, PAHs, and soot (Chuang et al. 1999; Roberts et al. 1996). Dust and soil particles smaller than $100-200\,\mu m$ in diameter may adhere to skin, clothing, and other objects and may be ingested through mouthing. Smaller particles (e.g., with diameters of $\sim 2-20\,\mu m$) may be resuspended into air, where they can be breathed into the upper respiratory system and lungs (Micallef et al. 1998; Thatcher and Layton 1995). Particles less than $10\,\mu m$ in diameter are inhalable and have a higher surface area per unit mass, which increases their inhalation toxicity. The concentrations of pesticides and PAHs in house dust are much higher on inhalable and respirable particles than on larger particles (Lewis et al. 1999).

Indoor pollution was ranked by the U.S. EPA as a high environmental risk 20 yr ago (GAO 1999; U.S. EPA 1987, 1990). The pollutants in house dust are an important component of indoor pollution. Meaningful progress has been made in monitoring and reducing lead dust exposure. Pollution monitoring and exposure analysis are required for efficient management of health risks and costs (Berube 2007a, b; Ott 2006). Recent research suggests that an economic analysis should be performed on the benefits of measuring and controlling the exposure of babies to all toxic compounds in house dust (Grandjean and Landrigan 2006; Lanphear et al. 2005; Louis et al. 2007; Maertens et al. 2008b; Roberts and Ott 2006).

The purpose of this article is to review and analyze the literature on monitoring and reduction of infant exposures to pollutants in house dust. We deal briefly with monitoring methods. We discuss concentrations in dust of a large variety of compounds including metals and persistent organic pollutants. We discuss methods of cleaning and ways to reduce exposure, particularly by home visits of trained volunteers. We conclude by suggesting hypotheses for further research related to cleaning and reduction of exposure by home visits.

2 Monitoring Pollutants in Surface and Deep Dust

Pollutants may be measured in environmental media (air, food, water, dust) and biological media (exhaled breath, blood, urine, hair, saliva). For the persistent pollutants of interest here, dust is a preferred sampling medium. Many pollutants have low vapor pressures and preferentially accumulate in dust, soil, or food. However, concentrations are generally higher in dust than in soil or food (U.S. EPA 2007).

Also, dust concentrations and loadings of pollutants show less variation over time than do air or urine concentrations (Egeghy et al. 2005).

2.1 Monitoring Methods

Measurement of dust levels on bare surfaces and in carpets is required to evaluate risks and control exposures. The concentration of pollutants in house dust (expressed as μg/g) can be measured by several methods including simply collecting and analyzing used vacuum cleaner bags from home (Colt et al. 1998). The dust loading on the surface of a carpet (in g/m^2) can be measured by weighing upright vacuum cleaner bags before and after vacuuming measured areas of a carpet eight times (Roberts et al. 1991b). However, it is difficult to compare such data collected with different vacuum cleaners, avoid cross contamination between samples, and extract small particles from the bag fabric.

Cassette samplers, rollers, wet wipes, and many other methods have been used to measure pollutant concentrations in house dust and in dislodgeable residues from carpets and upholstery (Farfel et al. 1994; Lanphear et al. 1995; Lioy et al. 1993; Hee et al. 1985; Roberts et al. 1991c; U.S. EPA 1989, 1995, 1996b). These methods will not be described here, but comprehensive reviews on them have been published by Lewis (2005), Lioy (2006), Pope et al. (1993), and the U.S. EPA (1997, 2007).

In 2001, the U.S. EPA developed a standard wipe method for measuring lead (Pb) loading on bare and carpeted surfaces. A standard of 40 μg Pb/ft^2 was established to protect small children, and this value has been used as a threshold for granting approval to rehabit old homes, after lead remediation or remodeling. Lead loading in carpet is among the best predictors of expected lead levels in blood of exposed toddlers (Davies et al. 1990). Moreover, carpet loading (μg of Pb/m^2) of pollutants correlates better with resuspended pollutant levels generated from activity on the carpet than does pollutant concentration (μg of Pb/g) (Lewis 2002; Roberts and Ott 2006).

A high-volume small-surface sampler (HVS3) described in ASTM International Standard Method D5438 (ASTM, 2007) was developed for the U.S. EPA in 1990 to assess risk from lead, pesticides, PAHs, and other pollutants in house dust on bare surfaces and carpets (Roberts et al. 1991a, c). The HVS3 allowed measurement of both concentration (μg/g) and loading (μg/m^2) of surface dust pollutants by using a cyclone and by controlling air flow and pressure drop across the nozzle. The cyclone allows collection of a large sample (up to 100 g) without any reduction in air flow. In 2002, the HVS3 was simplified to create the HVS4, which is less costly and easier to use and transport (Roberts et al. 2004). Several studies designed to measure children's exposure to pollutants (pesticides, lead, allergens, and PAHs) have used the HVS3 or HVS4 units (Roberts et al. 1999, 2004; U.S. EPA 2000; McCauley et al. 2001; Fenske et al. 2002; Bradman et al. 2007). One study compared HVS3 results to those using household vacuum cleaner bags and found little difference (Colt et al. 2008). Because the household vacuum cleaner bags presumably represented sample material collected over a long term than did the HVS3 samples, this provides further evidence that dust constitutes a stable matrix for pesticides.

2.2 Metals

Elements (metals) have been monitored in house dust in several studies since the 1990s (Rasmussen et al. 2001; Siefert et al. 2000). Examples of concentrations from a recent study in 78 California classrooms are shown in Table 1. Also shown in Table 1 are the U.S. EPA Region IX (which includes California) Preliminary Remediation Goals (PRGs) for 24 toxic elements in residential soils (Smucker 2004). These PRGs were selected, in part, to protect the health of infants. Exceeding these PRGs in residential soil at a Superfund site triggers a risk assessment, and these same PRGs may be used as standards for cleanup. The PRG values are usually set to provide an added margin of safety for cancer risk of less than one in one million, although cleanup standards may also be set for cancer risk rates of 1 in 100,000, or 1 in 10,000. The concentration of one toxic metal – arsenic – exceeded the PRG for the cancer endpoint in California classrooms.

Table 1 Metal concentrations (μg/g) in dust samples from California classrooms (N = 78)

Element	Preliminary remediation goals[a]	California classrooms[b]	
		Median	95th Percentile
Aluminum	76,000	47,500	60,100
Arsenic[c]	0.38 or 22	11.6	17.3
Cadmium	37	3.55	13.3
Cesium		0.24	0.70
Chromium	210	33.1	72.8
Cobalt	4,700	1.7	14
Copper	2,900	60.2	288
Iron	22,000	22,300	37,300
Lead	400	61.6	190
Magnesium		8,700	14,300
Manganese	1,800	316	417
Nickel	1,600	33.2	83.2
Palladium			4.03
Selenium	390	1.56	13.5
Strontium	47,000	139	235
Titanium	310,000	320	877
Vanadium	550	40	65
Zinc	23,000	980	2,020

[a] For residential soils, in ppm (Smucker 2004)
[b] CARB (2003)
[c] Arsenic PRG = 0.38 ppm for cancer endpoint, 22 ppm for noncancer

Toxicants may become concentrated in house dust reservoirs. The concentration of toxic metals in house dust may be from 2 to 32 times higher than the levels found in garden soil around the house (Rasmussen et al. 2001). The median concentration of mercury in house dust (1.61 µg/g) was 32 times that found in an Ottawa garden soil; this suggests that an indoor source of mercury also may exist. Moreover, the type of heat source used in houses affected mercury concentrations found in dust: electric (4.13 µg/g), gas (1.36 µg/g), and oil (1.39 µg/g). Mercury switches are one possible contamination source in homes utilizing such switches in electric heaters. However, it is difficult to pin exact sources down, because a wide variation in metal concentrations exists from one house to another and in the ratio of indoor to outdoor concentrations. There is a strong correlation between metal levels and concentrations of organic carbon in house dust (Rasmussen et al. 2001; Rasmussen 2004). The higher organic carbon content of urban fine house dust (27.5%), in relation to topsoil (4%), for particles less than 53 µm in diameter, may be one factor that increases indoor dust toxicity (Rasmussen 2004). However, metal concentrations in indoor dust cannot be predicted from outdoor soil levels (Rasmussen et al. 2001; Rasmussen 2004). The highest children's blood lead levels and lead loadings in carpets are associated with the following factors: home remodeling, paint removal, lack of an effective vacuum cleaner, infrequent cleaning, and peeling paint inside and outside of older houses (Davies et al. 1990; Roberts et al. 1991b, 1999, 2004; U.S. EPA 1997). Some research suggests that lead dust ingestion in young children may account for 1,000 times more exposure than inhalation (Roberts and Dickey 1995).

Egeghy et al. (2005) found that a single measurement of lead in blood or chlorpyrifos in house dust was sufficient for an estimate of average resident exposure. However, most other compound/media combinations required more measurements. Egeghy concluded that measurements in both biological fluids and dust were more consistent than those in indoor air.

U.S. EPA action to remove lead from U.S. gasoline resulted in a dramatic decline of lead levels in children's blood from 1976 to 1999. During this period, the median blood lead levels of children aged 5 and under dropped from 15 to 2.2 µg Pb/dl (U.S. EPA 2003). This reduction shows the potential of product reformulation for protecting children. However, one in three children, under the age of 6 yr, still live in older houses that retain a lead-based paint hazard (Clickner et al. 2001). Some 51% of 154 Seattle homes of Master Home Environmentalist (MHE) volunteers built before 1940 had house dust lead levels that exceeded the U.S. EPA PRGs of 400 µg/g (Roberts and Ott 2006; Smucker 2004).

2.3 Pesticides

Carpeting is a common dust reservoir and an efficient pesticide concentrator. Carpets collect soil particles tracked in from outdoors and collect settled dust from indoor air. Carpet-embedded dust, carpet fibers, backing, and padding can also absorb pesticides from liquid and aerosol sprays, gaseous pesticides in air within the home, or from vapors that intrude into the home from the crawlspace or basement. Typically, pesticide concentrations in vacuumable house dust are 10–100 times higher than those found in outdoor surface soil (Lewis et al. 1994; Simcox et al. 1995). Pesticide residues may persist for years in carpets, where they are protected from sunlight, rain, temperature extremes, and some microbial action.

Even if residents do not use indoor pesticides, track-in of lawn-applied pesticides can be of particular concern. The presence in house dust and indoor air of the herbicide 2,4-dichlorophenoxyacetic acid (2,4-D), the insecticide carbaryl, and the fungicide chlorothalonil, which are normally applied exclusively outside the home, implies that the pollutants have been transported from outdoors (Lewis et al. 1999).

An important community of interest for pesticide exposure is farmworkers, particularly migrant farmworkers, who may have children at greatly increased risk of exposure to pesticides transported into the home as residues on clothes or shoes. Several studies have focused on farmworker-family exposures to pesticides in house dust (Bradman et al. 2007; Freeman et al. 2004; Arcury et al. 2006; McCauley et al. 2001; Fenske et al. 2002; Thompson et al. 2008; Ward et al. 2006). Most of these studies found elevated levels in house dust and/or in urinary metabolites of the targeted agricultural pesticides.

U.S. EPA studies have shown that walking over pesticide-treated turf, as long as 1 wk after treatment, can result in transfer of residues to carpet dust in amounts proportional (3–4%) to the dislodgeable residues on the turf (Nishioka et al. 1996). Results from these studies indicated that 2,4-D residues, in the home from lawn applications, were measurably higher with active children and pets (Nishioka et al. 1999, 2001). Concentrations ($\mu g/m^3$) measured on 10-μm airborne particles were two to ten times higher than those on 2.5-μm particles, with concentrations declining on particles larger than 10 μm. Indoor residues persisted after lawn residues had dissipated.

A recent report (U.S. EPA 2007) has reviewed 13 studies, all carried out by the EPA between 1997 and 2001. Because the EPA report is both recent and quite complete, we will not repeat its findings here. Readers are encouraged to check this report both for the results of the reviewed studies, five of which involved more than 100 children each, and for a comparison of the many dust sampling methods employed in these studies.

The potential for exposure by ingestion versus inhalation depends on pesticide volatility. The primary route of exposure for infants and toddlers, who are often in close contact with the floor, is ingestion of either contaminated house dust or surface residues; intake of nonvolatile pesticides such as pyrethroids that are both abundant in carpet dust and widely used indoors and herbicides tracked in from lawns is a prominent source of such exposure (Lewis et al. 1999; Rudel et al. 2003; U.S. EPA 2000). For relatively volatile pesticides and those adhering to respirable particles inhalation may represent the primary exposure route. To illustrate this point, the authors of one study (Lewis et al. 2001) estimated that a young child's potential indoor exposure to diazinon (now cancelled for residential use) may be 50 times greater from inhalation than from ingestion of house dust at 100 mg/d [EPA's exposure guideline for infants and toddlers; U.S. EPA (1996a, 2002)].

Alternatively, ingestion by mouthing of diazinon residues on the hands of the children who participated in this study would have exceeded the inhalation exposure level by 2- to 3-fold.

Residues from pesticides discontinued long ago in the USA are still found in house dust. Chlordane (banned in 1988) was still detected in 38% of homes, whereas DDT (discontinued in 1972) was still found in 70% of house dust samples collected from 1998 to 2001 (Colt et al. 2004). The lower volatilities of DDT, DDE, carbaryl, and methoxychlor suggest that they will persist longer in house dust because lower amounts vaporize at ambient temperatures. Pesticides with higher vapor pressures may condense closer to the point of application in the winter and translocate to cooler climates in the summer, in a gas chromatographic or "grasshopper" effect (Riseborough 1990; Lioy 2006). Persistent pesticides with low vapor pressures may be transported long distances by foot and vehicular traffic, and on airborne fine particles, even to pristine snow-covered areas in the Western National Parks (Hageman et al. 2006; Kurtz 1990; Lewis and Lee 1976; Lewis et al. 1994, 1999; Lewis 2005; McConnell et al. 1998; Simcox et al. 1995).

In the USA, home use of common pesticides increased from 36 million kg (of active ingredients) in 1999 to 46 million kg in 2001; lawn-applied herbicides accounted for 71% of the total (U.S. EPA 2004). Not included in these figures are 27 million kg of nonconventional pesticides such as disinfectants, deodorizers, and insect repellants. Although accurate assessment of total exposure risks associated with pesticide use in and around the home remains difficult, it is clear that residents may be exposed to pesticide residues in untreated as well as treated areas of the home, and children may be exposed through intimate contact with both intentionally and incidentally contaminated surfaces.

2.4 PAHs and PCBs

Carpets are contaminated by and accumulate PAHs as well as pesticides. Table 2 presents the concentration distributions of prevalent pesticides, PAH, and PCB (polychlorinated biphenyl) congeners in house dust from subjects enrolled in a population-based case-control study of non-Hodgkins lymphoma (NHL), a large study in which semivolatile organic chemicals were measured in dust to investigate purported risk factors for NHL. Vacuum cleaner bag dust was analyzed if subjects had used their vacuum cleaner within the previous year and had owned at least half of their carpets for 5 yr or more. The median length of residence in homes was 20 yr for both cases and controls. Results indicated that NHL risk was elevated by 50% if any PCB congeners were detected; greater risk existed at higher PCB concentrations in dust, and there was evidence of greater effects for PCB 180 (Colt et al. 2005). NHL risk in men was elevated by 30%, if DDE was detected. Chlordane treatment of homes for termites elevated the resident's NHL risk by 30%, and NHL risk increased with increasing levels of chlordane in house dust (Colt et al. 2005, 2006). The only chemicals in dust found to elevate NHL risk had been banned for

Table 2 Concentration distributions of prevalent pesticides, polycyclic aromatic hydrocarbons (PAHs), and polychlorinated biphenyl (PCB) congeners in house dust collected from 616 homes in Detroit, MI, Los Angeles County, CA, Seattle, WA, and the State of Iowa, 1999–2001 and from 78 California classrooms

Pollutant	Preliminary remediation goals[a] ($\mu g/g$)	House dust[b]			Vapor pressure (kPa at [ ]°C)	Classroom dust[c]		
		% Detected	Median ($\mu g/g$)	95th Percentile ($\mu g/g$)		% Detected	Median ($\mu g/g$)	95th Percentile ($\mu g/g$)
Pesticide								
Carbaryl	6,100	35	0.050	3.01	4.1×10^{-8} [25][c]			
Chlordane	1.6	38	0.021	0.44	1.3×10^{-6} [25][c]			
Chlorpyrifos	180	68	0.108	3.53	2.7×10^{-6} [25][d]	97	0.308	1.906
2,4-D		78	0.30	7.01				
DDE	1.7	46	0.0237	0.20	4.2×10^{-7} [20][c]	54	0.008	0.052
DDT	1.7	70	0.09	1.6	2.5×10^{-8} [20][d]			
Diazinon	55	39	0.025	0.76	1.2×10^{-5} [25][d]	58	0.035	0.679
Methoxychlor	310	42	0.074	2.07	1.1×10^{-7} [25][f]			
Pentachlorophenol		87	0.37	3.18				
cis-Permethrin		72	0.33	20.9	2.5×10^{-9} [20][d]	99	0.256	1.870
trans-Permethrin		74	0.70	38.7	1.5×10^{-9} [20][d]	100	0.320	2.329
ortho-Phenylphenol		99	0.25	1.25	1.5×10^{-4} [25][h]	100	0.063	0.486
Propoxur		77	0.072	1.45	1.3×10^{-6} [20][d]			
PAHs								
Benz(*a*)anthracene	0.62	98	0.136	1.79	4.1×10^{-9} [25][h]	79	0.053	0.329
Benzo(*b*)fluoranthene	0.62	99	0.31	3.95	5.0×10^{-10} [25][h]			
Benzo(*k*)fluoranthene	6.2	91	0.099	1.10	5.2×10^{-11} [25][h]	80	0.057	0.378
Benzo(*a*)pyrene	0.06	96	0.154	2.39	3.0×10^{-10} [25][h]	59	0.054	0.306
Chrysene	62	99	0.27	2.84	4.0×10^{-9} [25][h]	93	0.149	0.678
Dibenz(*a,h*)anthracene	0.06	71	0.036	0.50	1.3×10^{-11} [25][h]	41	0.003	0.081
Indeno(1,2,3-*cd*)pyrene	0.62	96	0.161	2.39	1.0×10^{-11} [25][g]	68	0.049	0.357
PCB congeners								
PCB 105		18	<0.021	0.090				

(continued)

Table 2 (continued)

Pollutant	Preliminary remediation goals[a] (μg/g)	House dust[b]				Classroom dust[c]		
		% Detected	Median (μg/g)	95th Percentile (μg/g)	Vapor pressure (kPa at [ ]°C)	% Detected	Median (μg/g)	95th Percentile (μg/g)
PCB 138		33	<0.021	0.22				
PCB 153		40	<0.021	0.24				
PCB 170		14	<0.021	0.056				
PCB 180		26	<0.021	0.112				

[a] Smucker (2004)
[b] Usual detection limits are 0.021 μg/g for most listed pollutants; exceptions were 0.025 μg/g (methoxychlor), 0.027 μg/g (carbaryl), 0.055 μg/g (cis-permethrin), 0.060 μg/g (trans-permethrin), and 0.084 μg/g (2,4-D pentachlorophenol). Camann et al. (2002a, b), Colt et al. (2004)
[c] CARB (2003)
[d] Tomlin (2000)
[e] ATSDR (1993)
[f] Howard (1991)
[g] TOXNET (2007)
[h] Mackay et al. (1992)

at least 10 yr (PCBs, DDE, and chlordane). Considering the lack of recent use, detection of these agents in dust signified equal or greater exposure to residents for many years prior to dust sample collection. NHL risk was not found to be elevated for any of the current-use pesticides or PAHs listed in Table 2 (Colt et al. 2005, 2006; Hartge et al. 2005). Because carpet dust residues resulting from recent use are normally higher than those from the distant past and may obscure effects of aged residues, risks for cancers with a long latent period are more efficiently evaluated using dust measurements for chemicals for which there is no recent contribution.

A recent population-based case-control study in Northern California investigated risk of childhood leukemia in relation to exposure to organochlorine concentrations in HVS3-collected dust collected from carpets present in the home before the diagnosis/reference date (Ward et al. 2008). Leukemia risk was increased 2.8-fold in the highest versus lowest quartile for the sum of the six measured PCB congeners (105, 118, 138, 153, 170, 180), with a significant positive trend in leukemia risk as total PCB concentrations increased. Childhood leukemia risk was not associated with dust concentrations of chlordane, DDT, DDE, or pentachlorophenol (PCP).

The comparatively low PCB concentrations found in dust appear to be more strongly associated with incidence of both cancers (NHL and childhood leukemia) studied than are dust levels of any other semivolatile chemical measured. PCBs were used in fluorescent lighting fixtures and many other products manufactured before 1977 (ATSDR 2000), and a PCB-containing wood-floor finish, used in the 1950s and 1960s, can still be an important source of continual PCB exposure in some older homes (Rudel et al. 2008).

Seven PAHs (Table 2) have been classified as probable or known human carcinogens (IARC 1985, 2008). The frequency of detection of PAHs in carpet dust ranged from 71 to 99%, as compared with 35–77% for pesticides. The median and 95th percentile concentrations (µg/g) of chlordane and DDT, as well as five of the PAHs presented in Table 2, approached or exceeded the U.S. EPA PRGs. The median and 95th percentile concentrations of the known human carcinogen benzo(a)pyrene (BaP) exceeded the PRG (0.062 µg/g) by factors of 2.5 and 38.5, respectively (IARC 2008). The 90th percentile concentration for BaP, in Detroit house dust, was 4.8 µg/g, compared with 0.87, 0.88, and 0.21 µg/g in Iowa, Seattle, and Los Angeles, respectively (Camann et al. 2002a). The median BaP level in house dust in 120 homes on Cape Cod was 0.71 µg/g (Rudel et al. 2003).

The seven potentially carcinogenic PAHs presented in Table 2, and benzo(ghi) perylene, were detected in nearly all of the personal air samples collected from monitored pregnant minority women in New York City (Tonne et al. 2004). Prenatal exposure to these PAHs was associated with reduced fetal growth, increased risk of cognitive delay (Perera et al. 2006), and asthma (Miller et al. 2004). Chuang et al. (1999) suggest, in a pilot study, that low-income North Carolina adults received the following proportion of their exposure to PAHs that are rated by the U.S. EPA as probable human carcinogens (B_1 and B_2): diet (84%), air (9%), and dust ingestion (7%). Similarly, low-income children old enough to be toilet trained received 66% of their risk from diet, 10% from inhalation, and 24% from dust. Table 3 shows the

Table 3 Summary of PAH concentrations (μg/g) in house dust, entryway dust, and pathway soil from 24 low-income homes

Compound	House dust				Entryway dust				Pathway soil			
	Mean	Standard deviation	Minimum	Maximum	Mean	Standard deviation	Minimum	Maximum	Mean	Standard deviation	Minimum	Maximum
Naphthalene	0.33	0.85	0.02	4.30	0.11	0.26	0.01	1.31	0.01	0.01	<0.01	0.04
Acenaphthylene	0.08	0.06	0.01	0.27	0.04	0.06	<0.001	0.27	0.01	0.01	<0.001	0.03
Acenaphthene	0.05	0.04	<0.001	0.18	0.04	0.03	0.01	0.12	0.01	0.02	<0.001	0.08
Fluorene	0.12	0.24	0.02	1.22	0.04	0.03	0.01	0.06	0.01	0.01	<0.001	0.04
Phenanthrene	0.44	0.40	0.13	2.15	0.29	0.29	0.04	1.32	0.08	0.09	0.01	0.36
Anthracene	0.12	0.15	0.01	0.75	0.10	0.12	<0.001	0.40	0.03	0.05	<0.001	0.18
Fluoranthene	0.52	0.37	0.09	1.89	0.37	0.35	0.02	1.44	0.12	0.15	0.01	0.57
Pyrene	0.43	0.33	0.06	1.65	0.28	0.28	0.02	1.04	0.12	0.16	0.01	0.60
Benzo(a)anthracene[a]	0.22	0.17	0.04	0.69	0.15	0.14	0.01	0.52	0.06	0.09	<0.001	0.32
Chrysene[a]	0.39	0.47	0.05	2.41	0.20	0.19	0.02	0.74	0.02	0.02	<0.001	0.08
Cyclopenta[c,d]pyrene	0.08	0.05	0.02	0.22	0.05	0.03	<0.04	0.13	0.02	0.04	<0.001	0.20
Benzo(b and k)fluoranthenes[a]	0.55	0.32	0.17	1.34	0.36	0.28	0.04	0.08	0.12	0.15	0.01	0.47
Benzo(e)pyrene	0.26	0.17	0.05	0.75	0.20	0.15	0.02	0.54	0.05	0.06	<0.001	0.20
Benzo(a)pyrene[a]	0.23	0.15	0.07	0.63	0.15	0.13	0.01	0.41	0.06	0.01	<0.001	0.35
Indeno(1,2,3-c,d)pyrene[a]	0.23	0.18	0.05	0.70	0.16	0.13	0.01	0.42	0.05	0.07	<0.001	0.28
Dibenzol(a,h)anthracene[a]	0.10	0.09	0.02	0.41	0.05	0.05	0.01	0.15	0.02	0.03	<0.001	0.13
Benzo(g,h,i)perylene	0.25	0.16	0.08	0.61	0.17	0.13	0.01	0.37	0.05	0.06	<0.001	0.21
Coronene	0.13	0.11	0.04	0.50	0.10	0.11	0.01	0.51	0.03	0.04	<0.001	0.19
Sum of B2 PAH	1.73	1.25	0.46	5.98	1.08	0.86	0.13	2.87	0.40	0.51	0.02	1.77
Sum of target PAH	4.52	2.91	1.25	15.20	2.85	2.05	0.42	6.58	0.96	1.10	0.06	3.84

[a] Denotes B2 carcinogenic PAHs
Source: Adapted with permission from Chuang et al. (1999)

concentration of 18 PAHs in house dust, entryway dust, and pathway soil. The sum of the B_2 carcinogenic PAH concentrations in house dust exceeds four times the concentration found in the sum of the same compounds in entryway soil, which suggests preferential track-in of small particles (Roberts et al. 1996).

The low vapor pressures of the carcinogenic PAHs listed in Table 2 suggest that they will condense quickly after combustion, be deposited in soil, and be tracked into houses.

In cold climates downwind of large cities (London, New York, and Boston), where coal combusted at low temperatures has been burned for home heating, it is expected that stable PAHs such as BaP may accumulate in soil and house dust. In southeast England, the BaP in the "plow layer" (top 23 cm of soil) increased by a factor of more than 15 in the 100-yr period between 1880 and 1980; the total PAH burden in soil increased by a factor of 4. The flux rate for BaP was 0.36 mg/m² per year (Jones et al. 1989). Beyea et al. (2006) found a correlation between automotive traffic emissions to the air and soil concentrations of PAHs; air emissions were also correlated with the incidence of blood PAH–DNA adducts in women on Long Island, New York. They did not find a similar pattern with PAHs in house dust and suggest that indoor PAHs from cooking may have been an important source of indoor PAHs. The infant or adult diet may be the main source of PAH exposure, but the percentage of total exposure allocated to dust exposures for the carcinogenic PAHs may be three times higher (24% vs. 7%) among low-income children (Chuang et al. 1999).

The exposure of babies to BaP in house dust is worrisome, because BaP is a confirmed human carcinogen (IARC 2008) that has a possible impact on asthma and mental development (Miller et al. 2004; Perera et al. 2006). Moreover, BaP exists in up to 96% of homes, is detected at median concentrations that are 2.5 times higher than the PRG action level for residential soil at Superfund sites (Camann et al. 2002a), is persistent, and has increased its concentration in soil by a factor of 15 over a 100-yr period downwind from London (Jones et al. 1989). The largest source of BaP and other house-dust-associated toxicants may be from track-in of soil (Adgate et al. 1998; Chuang et al. 1999; Roberts and Ott 2006); this suggests that vacuuming, removal of shoes at the door, use of booties, and wiping shoes on a high-quality door mat may be good control strategies.

2.5 Phthalates

Although the health effects of infants' exposure to lead in house dust are well documented (Canfield et al. 2003; Lanphear et al. 2005; U.S. EPA 2003), less is known about infant exposures to semivolatile PAHs, phthalates, phenols, and PBDEs used as flame retardants, organotins, perfluorinated organics (PFOs), and dioxins in dust. Jones-Otazo et al. (2005), Stapleton et al. (2005), and Lorber (2008) suggest that most PBDE exposure to both children and adults comes from house dust.

Health-based PRGs have been established for residential soils at U.S. EPA Superfund sites for many EDCs such as pesticides and PAHs, but not for most phthalates, PBDEs, and PFOs, which are also EDCs (Moriwaki et al. 2003; Smucker 2004). PFOs are of particular concern because they are water soluble and difficult for humans to excrete.

Wildlife health effects are associated with exposure to EDCs that mimic or interfere with hormones, particularly the estrogen function, even though the causal link between exposure to EDCs and endocrine disruption may not be clear (Jobling and Tyler 2006). In the National Health and Nutrition Examination Survey (NHANES; 1999–2002), an association was found between abdominal obesity, insulin resistance, and several urinary phthalate metabolites in a cross section of US men (Stahlhut et al. 2007). An association was also found between a metabolite of the plasticizer DEHP (diethylhexylphthlate) in urine and reduced thyroid hormone in blood of 408 men in Boston (Meeker et al. 2007). The combined effect of EDC mixtures in house dust is unknown. A test was developed to measure the effect of mixtures of EDCs, or the total effective xenoestrogen burden, in the placenta at birth by Fernandez et al. (2007). These and other investigators suggest that the exposure to a mixture of EDCs during pregnancy may disturb the sexual development and increase the risk of hypospadias and cryptorchidism in male infants (Swan et al. 2005).

Rudel et al. (2003), in a study of 120 homes, found significant levels of EDCs in both indoor air and house dust as part of a case control breast cancer study on Cape Cod. Of 89 organic EDCs, 52 were detected in air and 66 in dust. Pesticide EDCs detected included 23 in air and 27 in dust. The most abundant EDC in indoor air was diethyl phthalate with a median concentration of 590 ng/m^3. This same phthalate was found in indoor air at a median concentration of 350 ng/m^3 in 125 Riverside, CA homes (Sheldon et al. 1992). DEHP was the most abundant EDC in house dust, in both the 120 Cape Cod homes and in 286 sampled German homes, with median concentrations of 340 and 740 µg/g, respectively (Table 4). These and other phthalates found in dust came from plasticizers and emulsifiers in plastics, food packaging, building materials, and personal care products. The concentrations detected exceeded the PRGs or other health-based guidelines for 15 compounds in air or dust, but there were no such guidelines for 28 of the 89 EDCs. In addition to endocrine and reproductive effects, phthalates have been implicated in asthma, which may be related to their ability to induce oxidative stress (Bornehag et al. 2004; Jaakkola et al. 2006; Kolarik et al. 2008).

2.6 PBDEs

A rise in Canadian and US blood and breast milk PBDE concentrations, in the 1990s and 2000s, suggests that exposure to PBDEs in house dust was increasing during this period (Stapleton et al. 2005). House dust is the source of an estimated 82% of

Table 4 Levels of PAHs, pesticides, phthalates, phenols, polybrominated diphenyl ethers (PBDEs), alkylphenols, and alkylphenol ethoxylates in house dust from 120 Cape Cod homes and 196 or 286 German homes

Pollutant	Preliminary remediation goals[a] (μg/g)	Cape Cod house dust[b]			German house dust[c]	
		% Detected	Median (μg/g)	Maximum (μg/g)	Median (μg/g)	95th Percentile (μg/g)
PAHs						
Pyrene	2,300	96	1.33	39.8		
Benz(a)anthracene	0.62	76	4.99	10.0		
Benzo(a)pyrene	0.062	85	0.712	18.1		
Pesticides					*196 Homes*	
4,4-DDT	1.7	65	0.279	9.61	0.31	4.2
Chlorpyrifos	180	18	RL[d]	228	<0.1	0.63
α Chlordane	1.6	39	RL	9.97		
Methoxychlor	310	54	0.240	12.9	0.92	27
Pentachlorophenol	3.0	86	0.793	7.96	0.95	8.0
Phthalates					*286 Homes*	
Dicyclohexyl phthalate		77	1.88	62.7		
Diethyl phthalate	49,000	89	4.98	111		
Di-n-butyl phthalate		98	20.1	352	49	240
Benzylbutyl phthalate	12,000	100	45.4	1,310	31	320
Di-iso-butyl phthalate		95	1.91	39.1	34	130
Diethylhexyl phthalate (DEHP)		100	340	7,700	740	2,600
Bis(2-ethylhexyl) adipate		100	5.97	391.5		
Phenols						
Bisphenol A		86	0.821	17.6	3.4	9.2
2,4-dihydroxybenzophenone		63	0.515	9.36		
PBDEs						
PBDE 47		45	RL	9.86		
PBDE 99		55	0.304	22.5		

(continued)

Table 2 (continued)

Pollutant	Preliminary remediation goals[a] (µg/g)	Cape Cod house dust[b]			German house dust[c]	
		% Detected	Median (µg/g)	Maximum (µg/g)	Median (µg/g)	95th Percentile (µg/g)
PBDE 100		20	RL	3.4		
Alkylphenols and alkylphenol ethoxylates						
4-Nonylphenol		80	2.58	8.68		
Nonylphenol monoethoxylate		86	3.36	15.6		
Nonylphenol diethoxylate		86	5.33	49.3		
Octylphenol monoethoxylate		50	0.13	1.99		
Octylphenol diethoxylate		69	0.306	2.12		

[a] U.S. EPA preliminary remediation goals (PRGs) for superfund residential soils, in ppm (Smucker 2004)
[b] Source: Rudel et al. (2003). <150 µm in diameter from thimble on vacuum wand
[c] Source: Butte and Heinzow (2002). <63 µm in diameter from occupant vacuum bags
[d] RL = reporting limit

adult PBDE exposure; the percentage is even higher for children (Lorber 2008). The estimated PBDE dose received by children 1–5 yr of age is six times higher than for adults (49.3 vs. 7.7 ng/kg/d), because of their lower body weight and higher dust intake. Lorber (2008) suggests that the higher concentration of PBDEs in blood and breast milk in the USA versus Europe is related to the tenfold higher concentrations found in US house dust. The average house dust concentrations of PBDE 47, PBDE 99, and PBDE 100 in 89 Massachusetts homes (Rudel et al. 2003) were 700, 1,290, and 170 ng/g dry mass, respectively, as compared with concentrations for the respective PBDEs in 40 German homes of 20, 180, and 20 ng/g (Stapleton et al. 2005).

Two of the three commercial PBDE mixtures, PentaBDE and OctaBDE, have been voluntarily withdrawn or banned from use in some parts of the world. However, a recent study has shown that residues of at least five of the alternative or new brominated fire retardants are showing up in Boston homes (Stapleton et al. 2008).

2.7 Phenols and Alkylphenols

Some phenols are also reported to be estrogenic EDCs (Rudel et al. 2003). The alkylphenols are impurities or degradation products of alkylphenol polyethoxylates used in detergents, personal care products, and as inert ingredients in pesticide formulations (Rudel et al. 2003). Table 4 shows levels of various PAHs, pesticides, phthalates, phenols, PBDEs, alkylphenols, and alkylphenol ethoxylates found in 120 house-dust samples collected from Cape Cod, four phthalates from 196 German homes, and four pesticides and bisphenol A (BPA) in a separate group of 286 German homes (Butte and Heinzow 2002; Rudel et al. 2003). The size of the house dust particles analyzed in the Cape Cod study was <150 μm in diameter, whereas particle size in the German study was <63 μm.

Wilson et al. (2007) found BPA in house dust in 25% of 119 North Carolina homes (95th percentile=0.236 μg/g), and in 47% of 116 Ohio homes (95th percentile=0.141 μg/g). The median values were below detectable levels. In the same study, they found PCP in house dust in 92% of 121 North Carolina homes (median 0.060 μg/g, 95th percentile 0.492 μg/g) and in 94% of 119 Ohio homes (median 0.060 μg/g, 95th percentile 0.345 μg/g). The authors analyzed children's exposure to BPA and PCP in homes and day care centers in food, outdoor air, indoor air, soil, and house dust. The potential incidence of exposure to BPA was 99% in both states, primarily from dietary ingestion. The authors estimated that the proportion of those exposed to PCP came primarily from inhalation: 78% in NC and 90% in Ohio (Wilson et al. 2007).

Camann et al. (2005) studied dust (particle size< 150 μm) taken from vacuum cleaner bags from ten homes in each of seven states (CA, MA, ME, MI, NY, OR, WA). The mean cumulative concentration of targeted EDCs found ranked by descending percent detected was as follows: five diester phthalates (mainly DEHP), 424 μg/g (90%); six alkylphenols and their ethoxylates, 26.7 μg/g (5.6%); nine pesticides, 12.7 μg/g (2.7%); six PBDEs, 9.1 μg/g (1.9%); four organotins, 0.65 μg/g (0.1%); and perfluorinated chemicals, 0.50 μg/g (0.1%).

2.8 Dust Mites, Mold, Other Allergens, Viruses, and Bacteria

Asthma rates have more than doubled in the USA since 1980, and asthma is the most common chronic illness, cause of hospitalization, and school absence of children in the USA (Akinibami 2006; Institute of Medicine 2000). Dust is both a reservoir and a vehicle that enhances human contact with asthma triggers such as dust mite allergen, mold, animal dander, cockroach and rodent allergens, PAHs, smoke deposits, and many other compounds in house dust that irritate the lungs and mucous membranes. House dust can be a trigger for asthma and allergy attacks.

There is sufficient evidence that exposure to dust mites can cause asthma in susceptible children (Institute of Medicine 2000). Sensitization of children to dust mites is one of the strongest risk factors for persistent asthma in adults (Sears et al. 2003). The highest concentration of dust mite antigen in dust is found in beds that maintain a mild temperature, contain moisture, and have a copious supply of skin scales. These conditions are optimum for dust mite growth and reproduction. When a baby or adult moves in bed and rubs against bedclothes the dust mite antigen (fecal pellets) are resuspended from the sheets or pillow and can be inhaled. The fabric in allergy control covers allow air to pass through, but the tight weave does not allow penetration of dust mites or their pellets. Although dust mites have a diameter similar to the size of a period on this page, they are rarely seen because they are transparent and flee from light (Arlian and Morgan 2003). Removal of dust from a bedroom carpet or removal of the carpet from a bedroom are important ways to reduce exposure to dust allergens (Institute of Medicine 2000). It is much easier to remove skin scales and dust mite allergens from a bare floor than from a carpet (Roberts et al. 1999).

The presence of mold in a home can double the risk of childhood asthma (Jaakkola et al. 2005). Mold commonly collects in house dust (Roberts et al. 1999). Mold and dampness can have the same large effect as does secondary tobacco smoke on child asthma, allergies, bronchitis, and other health problems, and these have a greater effect on children than on adults (Brunekreef et al. 1989; Institute of Medicine 2000). Children are more easily sensitized to environmental exposures (mold and other allergens) than are adults (Selgrade et al. 2006).

Dust is a source of bacteria (Roberts et al. 1999) and viruses in saliva and tracked-in bird, rodent, cat, and dog feces (Benson 1985; McCaustland et al. 1982). Endotoxin, produced by Gram-negative bacteria, is closely associated with asthma (Thorne et al. 2005).

2.9 Hygiene Hypothesis

Results of some studies suggest that exposure to higher levels of endotoxin from indoor Gram-negative bacteria or other infectious agents in house dust or during personal contact in early life may protect against allergen sensitization that may lead to allergies and asthma (Gereda et al. 2000; Zeldin et al. 2006).

Daily exposure to farm animals, early childhood infection from being raised in a large family, or attendance at daycare appear to protect against asthma. These observations have led to the "hygiene hypothesis," the idea that early exposure to infections may strengthen the immune system. Some families may assume that more house dust and less cleaning are good for their children. However, it was reported from a study of 4,000 school-aged children in Southern California that exposure during the first year of life to wood or oil smoke, cockroaches, herbicides or other pesticides, farm dust, or farm animals was associated with prospects for an asthma diagnosis before age 5 (Salam et al. 2003). In addition, some risk factors for asthma such as respiratory viral infections, obesity, pesticide exposure, and living in an inner-city community do not fit with the hygiene hypothesis (Zeldin et al. 2006).

There are serious flaws in the concept of exposing infants and others to low doses of pollutants in hopes of stimulating a protective response (Thayer et al. 2005). Exposing children to PAHs or other carcinogens, pesticides, endocrine disruptors, and neurotoxins that coexist with allergens in house dust cannot be justified by the hygiene hypothesis. It is prudent to reduce infant exposure to house dust by appropriate house cleaning and hand washing, while continuing to search for a safe method for protecting babies against sensitization to allergens.

3 Cleaning Practices, Carpets, and Safer Cleaning Products

3.1 Cleaning

Effective cleaning of homes and buildings can reduce dust and indoor air exposures to particles, lead, bacteria, allergens, and other pollutants (Franke et al. 1997; Gereda et al. 2000; Krieger et al. 2005; Lioy 2006; Roberts 2007). Cleaning helps protect the health of infants and adults. Effective home cleaning also encompasses preventive measures such as removing shoes at the entrance door to better control dust track-in. Other effective measures include providing commercial-grade door mats and regularly monitoring and extracting removable dust from carpets, plush furniture, bare floors, and other surfaces (Roberts 2007). Effective cleaning also improves indoor air quality. A year-long study, designed to measure the benefits of cleaning, in a large day care center (without any known indoor air problems) reduced indoor airborne dust by 58%, and total bacteria, Gram-negative bacteria, and fungi (mold) by 40%, 88%, and 61%, respectively (Franke et al. 1997). These pollutant levels were measured for 5 mon before and 7 mon after the cleaning program was introduced. More frequent and thorough cleaning is required in damp and warm climates to control dust and bacterial growth, dust mites, and mold that utilize the moisture and nutrients in dust to grow. Higher levels of cleaning are also required in houses built before 1978, when lead was phased out of paint; special attention is needed for houses built before 1940 because higher lead levels are normally present.

Most pollutants in dust are tracked in from outdoors (Adgate et al. 1998). Preventing dust track-in is easier than removing dust from a carpet. Shoe removal at the entrance is the best way to prevent track-in; simply wiping one's feet twice on a high-quality door mat may be 75% as effective as shoe removal (Roberts et al. 1991b). Commercial-grade door mats such as the "Twister" mat, used at many store entrances, are effective. However, such mats cost two to three times as much as most home door mats and are difficult to find in retail stores. They can be specially ordered at large hardware stores or purchased from commercial mat companies found in the yellow pages.

3.2 Carpets and Alternatives

Some carpets are much easier to clean than others. Flat and level loop carpets, often found in office buildings, are the easiest to clean. Short plush carpets are the next easiest to clean. Deep plush and shag carpets are the most difficult to clean. Spilled liquids may initiate mold growth in or under a carpet (without a water barrier on the pad) that is not dried within 24 hr. In 2005, the cost of installing a Powerbond carpet in Seattle was $3.50/ft^2. In 2008, the cost of installing alternative floor coverings in Seattle was as follows (per square foot): Berber carpet $1–$4, wood laminate $6–$8, tile $9–$11, and hardwood $8–$12. Wood laminate floors are popular but are not advertised as waterproof.

Carpets have advantages in reducing noise, are softer to walk and crawl on, may reduce the severity of falls, and have a lower cost than bare floors. Throw rugs and small area rugs present a hazard on bare floors and are a source of falls for both children and adults (Braun 2003). Larger carpets can provide greater traction than bare floors and may reduce injury in the event of a fall (Bunterngchit et al. 2000; Maki and Fernie 1990). However, replacing carpets with bare wood floors reduces cleaning time and also reduces exposure to dust and associated pollutants found in carpets and carpet backing. Older carpets tend to have more dust and health risks (Kim and Ferguson 1993; Roberts and Ott 2006). Hardwood and tile floors are more expensive than carpets, but last longer. Replacing carpets is usually not an option for most people who rent, but they can improve cleaning and hand washing. They can remove deep dust by using a vacuum cleaner with a dirt detector that electronically affirms when all removable dust has been vacuumed.

Plush furniture collects dust in the same way that plush carpets do and must be cleaned with a vacuum cleaner attachment or hand vacuum cleaner that has a rotating brush. Covering plush furniture with a washable fabric cover is one way to reduce exposure to accumulated dust.

3.3 Vacuum Cleaners

The carpets found in most homes collect dust and pollutants as they age (Kim and Ferguson 1993; Roberts et al. 1991a, 1999, 2004). Normal vacuuming removes surface dust but allows deep dust to accumulate, which may become surface and airborne dust after activity on the carpet (Roberts and Ott 2006). The total removable dust in a carpet is the sum of the surface and extractable deep dust. In two studies, the total removable dust (median 26.3 and 63.2 g/m^2, respectively) in samples of Seattle carpets more than 10-yr old exceeded the surface dust (median 1.3 and 2.9 g/m^2) by a factor of 20 or more (Roberts et al. 1999, 2004). Less than 20% of the lead, PAHs, and pesticides are usually found in the dust that can be extracted from older carpets by vacuuming, whereas, more than 80% of lead and 90% of PAHs and pesticides remain in the carpet fibers, backing, and pad (Roberts and Dickey 1995; U.S. EPA 2000). Pollutants with a higher vapor pressure (including chlorpyrifos, diazinon, and some PAHs) are unlikely to be removed from carpets by cleaning; these pollutants may continue to degrade indoor air quality and be deposited on room surfaces (Lioy 2006; U.S. EPA 2000). The source of the persistent concentrations of chlorpyrifos in indoor air, following many air exchanges, could not be explained until this large source was discovered.

The development of vacuum cleaners fitted with dirt sensors improved cleaning performance, because these units indicate when dust is still being collected from a carpet; such sensors allow monitoring of the effectiveness of extracting removable dust from a carpet. A Hoover vacuum cleaner, equipped with a dirt finder sensor using red and green indicator lights, was used to develop a standard "three-spot test method" to estimate the deep dust content and the time to remove it from a carpet (Roberts et al. 2004).

The three-spot test provides a measure of the time (in seconds) for the vacuum cleaner indicator lights to turn from red to green, when three locations are vacuumed 3 ft apart and at least 4 ft from an entrance door. The vacuum cleaner is held motionless on each spot until the light turns green. Achieving three green lights in less than 11 sec indicates that the carpet is relatively clean and retains less than 10 g/m^2 of removable dust. If the three-spot test takes 40 sec, the carpet may contain 80 g/m^2 of removable dust. A time of 6 sec on the three-spot test is an achievable goal. In contrast, a clean bare floor passes the three-spot test in less than 1 sec, with green light showing immediately everywhere. In most cases, the three-spot test can be performed in 1 min by families and professional cleaners to determine when the carpet is clean and all the removable dust has been extracted.

When deep dust was removed from 10-yr old carpets in daily use, the surface dust was reduced by 84–99% (measured 1 wk later with the HVS3; Roberts et al. 2004). Camann and Buckley (1994) collected data from 362 Midwest homes and estimated the median loading of surface dust on sampled carpets to be 1.4 g/m^2. This loading level can be reduced to less than 0.1 g/m^2, when deep dust is removed (Roberts et al. 2004).

The three-spot test correlates with, and gives a better indication of the total removable dust than does measuring surface dust (Roberts et al. 2004). Performing a walk-through survey and questionnaire by a trained outreach worker and conducting a three-spot test for dust in carpets is a low cost and efficient way to monitor dust and other exposures in a home (Roberts and Ott 2006).

Not all vacuum cleaners are equally effective. A canister vacuum cleaner without a power head can allow up to 400 g/m^2 of removable dust to collect in area rugs or plush carpets. Vacuum cleaners that utilize a power brush are two to six times more effective in carpet cleaning than are those without this device. A vacuum cleaner with a power head removes approximately 35–55% of added dust from a plush carpet and 70–80% from a flat or level loop carpet (personal communication: John Balough, The Hoover Company, 1988). The canister cleaner without a power head removes only 4% of resident dust from a shag rug, 10% from a plush carpet, and 40–50% from flat carpets. Ten to 25 times as much dust, lead, pesticides, and other pollutants may remain in an old rug than is removed by normal vacuuming (Roberts et al. 1999, 2004; U.S. EPA 2000).

Nearly any type of vacuum cleaner will remove all dust from a bare floor and is far superior to cleaning (sweeping) with a broom. Brooms may cause more particles to become airborne while leaving many particles behind. Wet mopping with a detergent and rinsing will help remove soot from a bare floor and is an option for poor families, who cannot afford a working vacuum cleaner. Area rugs on bare floors act as magnets for dust and should be cleaned by vacuuming with a power brush. It takes longer to clean a rug near entrance doors and in high-traffic areas because there is more dust to remove. If all the removable deep dust is not vacuumed from a carpet, in approximately 5% of cases, one may end up with more dust and pollutants on the carpet surface than was originally present (Roberts and Ott 2006). Vacuuming in two perpendicular directions may reduce the time required to get the green light on the dirt finder (personal communication: Debra Tucker, The Hoover Company, 2003). Vacuum cleaners will not work effectively if the bag is full and if the belt or brush is worn or broken. Bags should be changed when they are half full. When vacuum cleaners are not maintained, they lose their effectiveness and should be serviced.

Dusting with a damp cloth is one of the best ways to clean furniture, walls, and woodwork. Dusting should be done after vacuuming to remove any dust stirred up by this activity. The cloth must be changed or cleaned every 10 min, depending on how much dust is present, to maintain efficiency. Once the removable deep dust is vacuumed from the carpet, less dust will accumulate and be visible on woodwork and other surfaces.

Annual hot water-extraction cleaning of carpets helps to restore color and remove the soot that clings to carpet fibers that cannot be removed by vacuuming. Shampooing carpets with a detergent is often used to remove soot and stains, but increases the risk for mold unless the carpet and pad dry within 24 hr. Detergent residues left in the carpet after shampooing may contain pollutants. Hot water extraction does not require use of any detergent. Wet vacuuming is not as effective as dry vacuuming in removing dust (Lewis 2002). Removal of deep dust by dry vacuuming prior to commercial truck-mounted hot water extraction (or steam cleaning) provides one option for restoring color to a carpet, with less chance of leaving a damp carpet, dust, or increased surface lead. An all-purpose detergent is recommended for removing lead from surfaces (Lewis et al. 2006). Sending area rugs to a professional cleaner that features inversion, beating, vacuuming, immersion in a detergent solution, and rinsing is expensive but may remove pollutants that would otherwise remain in the fibers and backing.

3.4 Hand Washing

Hand washing with soap and water or with an alcohol-based fluid is the best way to stop the spread of infectious diseases and prevent transfer of pollutants in dust from surfaces to hands, food, mouth, and eyes. Hand sanitizing is widely used by the Group Health medical facilities in the Northwest USA. It has been found to be effective in both hospitals and homes, partly because it is more convenient and much quicker for busy parents and medical personnel (Sandora et al. 2005). Hand washing can reduce the 3.5 million annual deaths of children that result from diarrhea and acute lower respiratory tract infections (Luby et al. 2005; Nenstiel et al. 1997). Infectious diseases are the leading cause of death worldwide, and the third leading cause of death in the USA. Such diseases are also a leading cause of childhood illness (Nenstiel et al. 1997). Because infants are more vulnerable to infectious disease and ten times more vulnerable to pollutants on dust than are adults (U.S. EPA 2003), they have the most to gain from caregiver hand washing. Under normal circumstances, using regular soap without antimicrobial ingredients should be sufficient (Larson et al. 2004).

Infants can be protected by caring adults, who wash a baby's hands often, and who wash their own hands after cleaning, changing a diaper, sneezing into their hands, gardening, or going to the bathroom, and before preparing food, eating, or handling a baby (Luby et al. 2005). Supplying the infant with a clean pacifier is a good way to reduce finger sucking and the associated ingestion of dust and bacteria. Frequent washing of toys or other objects that a child puts in its mouth and putting a clean sheet down before you put an infant on the carpet also reduces dust and bacteria exposure. Good personal hygiene that includes hand washing, clean clothes, and effective house cleaning is a proven way to protect workers and infants exposed to dust contaminated with lead, arsenic, and other pollutants (OSHA, 1987; Davies et al. 1990).

3.5 Safer Cleaning Products

While cleaning the home is important, the choice of cleaning products and safe storage of them are also necessary to protect children's health. Cleaning products are frequently involved in home exposures reported to poison centers, with children under the age of 6 representing about half of all cases (Watson et al. 2005). In 2004, the percentage of moderate to major health outcomes, including death, reported from exposure to corrosive products was 14.03% (alkali drain cleaners), 14.33% (alkali oven cleaners), 11.14% (hydrofluoric acid rust removers), and 8.30% (acid toilet bowl cleaners). In comparison, for all other cleaning products, the percentage was only 3.56%. Corrosive products can be easily identified by the presence of the signal word "Danger" on product labels, in conjunction with the word "corrosive." A few solvent-based cleaning products such as furniture polishes and metal polishes also carry a "Danger" signal word because of their aspiration hazard if ingested. Sodium hypochlorite and ammonia are respiratory irritants, and products containing sufficient concentrations of these ingredients can irritate the lungs when used with inadequate ventilation. Strong respiratory irritants can also be produced if hypochlorite is mixed with ammonia or strong acids.

Choosing the least-toxic cleaning (and pest control) product to use in the home is difficult. This is because few ingredients are listed on product labels, and label warnings are primarily oriented toward acute exposures (Dickey 2005). Avoiding products with "Danger" or "Poison" signal words, in favor of those with "Caution" or no signal words, can remove the most acutely toxic products from homes, but does not guarantee that all ingredients are benign. All cleaning products should be securely stored out of reach of children.

4 Reducing Exposure and Health Costs with Home Visits

Environmental exposures are associated with increased rates of disease and health costs. Reducing exposures, when combined with health-improving behaviors, has the potential to reduce health costs by more than 30% (Eyre et al. 2004; Louis et al. 2007). Incentives are being used to improve health-related behaviors and can be used to reduce exposures of infants. Resources used to protect infants from exposures may have a much greater impact on health costs, learning, and productivity than the same resources used to protect older people, because babies are more vulnerable to exposures (U.S. EPA 2003). Such differences can be taken into account when doing exposure analysis (Ott 2006).

4.1 Home Surveys

The Home Environmental Assessment List (HEAL™©) used in the MHE Program of the American Lung Association of Washington (ALAW), together with the three-spot test mentioned previously, is a cost-effective way to develop a family action plan to reduce dust and total infant exposure. The MHE program helps people reduce their exposures to dust and indoor air by using trained volunteers to do a walk-through survey in the home with a family member Dickey (2007). Some 65 questions are asked that relate to exposures in the home, which helps the family establish priorities for improvement. The topic areas covered in the training and HEAL™© include lead, dust, indoor air, mold, moisture, biological contaminants, household chemicals, ventilation, tobacco cessation, asthma, allergies, and behavior change. At the end of the HEAL™© survey the family and the volunteer negotiate the top three actions the family is willing to do within 6 wk. Volunteers make a follow-up phone call after 2 wk to answer any questions and to reinforce the suggested changes.

The MHE program has spread from the ALAW of Seattle to Tacoma, WA; Yakima, WA; Wenatchee, WA; Colville, WA; Okanagan, WA; Olympia, WA;

Portland, OR; Boise, ID; New York, NY; Washington, DC; Fresno, CA; San Jose, CA; Bismark, ND; Providence, RI; San Antonio, TX; and Tulsa, OK. More than 800 volunteers have been trained in Seattle as of May 2008.

Although information is essential to induce families to make the changes required to reduce indoor exposures in a timely manner, it is insufficient when used alone (Krieger et al. 2005; Wu and Takaro 2007). Other factors that may help an individual or family make permanent changes include home surveys, support from a doctor, nurse, outreach worker, or friend. Laws and regulations that reduce environmental emissions and exposures to lead and other pollutants, in outside air, have little effect on in-home exposures that require changes in cleaning methods, home maintenance, ventilation, and use of consumer products. The role of MHEs in home exposures may, therefore, have a favorable impact. Trained MHEs and community health workers (CHWs) can reach where regulations cannot. MHEs can give motivational interviews that empower a family by meeting their unique needs and advise them on many low or no cost actions a family can take to improve cleaning, hand washing, ventilation, and selection of home chemicals.

Taking action to create a healthier home will provide important gains in comfort, productivity, and quality of life. Those individuals and families who suffer from asthma and allergies, or who are concerned about the indoor risks to babies and sensitive individuals, will be most willing to seek help in assessing and controlling indoor exposures. Other sensitive groups are the elderly, those ill with cancer, heart disease, AIDS, and poor immune systems, or those who have "sick-building"-associated health problems.

4.2 Reducing Asthma and Health Costs

Leung et al. (1997) and Primomo et al. (2006) have evaluated the effectiveness of the MHE program in Seattle and Tacoma, respectively. Leung found changes in family cleaning behavior after a home assessment by a MHE. Primomo found that 60 of 64 families (94%) made at least one change in cleaning behavior. Also, 34 out of 39 families (or 87%) with an asthmatic member reported an improvement in the condition of the sufferer after a HEAL™ evaluation and changes to cleaning behavior. The HEAL™ evaluation, by a trained community volunteer, is an effective way to increase knowledge about the home environment, asthma, and allergy triggers. Families who evaluate their home situation with a HEAL™ assessment tend to engage in behavioral changes and perceive improvements in asthma and allergy symptoms. Improvement in dust control is one of the most frequent changes after a HEAL™ assessment (personal communication: Maruyama, ALAW, 2005).

As part of the Seattle-King County Healthy Home Project I, Krieger et al. (2005) evaluated the benefit of environmental intervention by CHWs in reducing exposure to asthma triggers in the homes of 274 low-income asthmatic children. The families in the high-intensity group were given allergy control bedding covers, vacuum cleaners (with a dirt finder), high-quality doormats, and cleaning kits.

The Hoover Company provided the premium vacuum cleaners at cost, but such vacuum cleaners could be purchased at retail for $125, in 2008. The CHWs made five to nine visits over 1 yr focusing on asthma trigger reduction. The three-spot test (Roberts et al. 2004) was used to evaluate the effectiveness of carpet dust control and was performed at the beginning and end of the program. In 74 homes for which data were available, the average time to get three green lights on the dirt finder-equipped vacuum cleaners dropped 61%, from 59 to 23 sec, and the median time dropped 23%, from 15 to 11.5 s (Takaro et al. 2004). This test was attractive to both the CHWs and caregivers because it could be performed in less than 1 min, tended to reinforce effective cleaning, and the client could see immediate results.

Krieger et al. (2005) also documented reductions in exposures to asthma triggers, the severity of asthma, and health care costs. This trial was randomized and controlled and designed to compare results of a "high-intensity" intervention (multivisits and equipment used) with that of a "lower intensity" intervention (single visit, no special equipment used). In the higher intensity group, the incidence of nighttime symptoms, during a 2-wk period, dropped from 4.5 to 0.9 (80%); the incidence of daytime symptoms dropped from 8.1 to 3.4 (58%) and that of daytime symptoms with activity limitations dropped from 5.6 to 1.5 (73%). The proportion of participants using urgent health care services (unscheduled doctor visits and the emergency room) dropped by 15%. There was also a reduction in the use of medicines. The frequency of missed school days dropped from 31.1% to 12.1%. The percent of parents who missed work dropped from 13.1% to 11.2%. There was a significant improvement in the quality of life of the caregiver as measured by the Juniper Scale (Juniper et al. 1996).

The program cost per high intensity client was $1,345/yr, with an estimated annual medical care savings of $1,205–$2,001. This project also reduced the total exposure of the children to lead, pesticides, EDCs, and carcinogens as well as asthma triggers in dust. The lower intensity group received only allergy control mattress and pillow covers, plus one visit, at an annual cost of $222, and savings of $1,050 to $1,786, annually. This study did not collect follow-up data on both groups, but utilization of urgent care remained low among the high-intensity group for at least 6 months following the intervention. If one assumes this lower cost of urgent care, observed 6 mon after exit among the high-intensity group, persisted for 4 yr, the high-intensity intervention would be less costly, relative to the low-intensity intervention. The savings per child, discounted at 3%/yr, would range from $1,316 to $1,849 for 4 yr. Wu and Takaro (2007) demonstrated the cost-effectiveness of using a combination of environmental interventions including the use of CHWs for childhood asthma.

In 2007, the Washington State Legislature approved funding to help 700 low-income Medicaid children in King County, who had moderate to severe asthma, over a period of 4 yr. The same methods described by Krieger et al. (2005) will be used and are expected to cost $789 per child, with up to three home visits from trained CHWs, who will be supervised by nurses and doctors. Savings in health costs averaging $2,200 per child per year are estimated and result from fewer visits to hospitals, emergency rooms, and doctors. Standards for selection, training, and supervision of CHWs can be developed or adjusted to aid in the transfer of this program to the 38,000 other children in the Washington State or to the large number of children in the entire USA who have poorly controlled (moderate to severe) asthma (Dilley et al. 2005). It is estimated that 33% of Medicaid children with this level of asthma severity are less than 6 yr of age (personal communication: J. Krieger, 2007). The integrated home visits designed to improve the asthma and early learning of this large group of low-income children can be financed in part by the reduction in medical costs. Such children are expected to miss fewer days of school. They will be able to play, learn, work, and live better with less asthma.

The National Cooperative Inner-City Asthma Study (NCICAS) of 1,033 low-income asthmatic children was performed in Seattle and seven other cities (Morgan et al. 2004). These children were given allergy control pillow and mattress covers and the family provided education and social support in controlling asthma triggers in the home. NCICAS also reported a reduction in asthma severity. This study found that it was cost-effective to spend an average of $245 per year per child to achieve a $9.20 saving per symptom-free day gained. The sicker the child the greater the health cost savings. Managing exposure to asthma triggers, rather than the associated disease, could result in important social and financial benefits (Sullivan et al. 2002; Kattan et al. 2005).

Other studies also show the benefit of home remediation to reduce asthma triggers (Wu and Takaro 2007; Kercsmar et al. 2006). As more people become aware of this information, it is likely that families will ask for more assistance to reduce their everyday exposure to pollutants. Government as well as public and private groups have an important role to play in disseminating these research findings to the public on how to reduce their exposure, risks, and costs. Involving the community in planning and at the start of an environmental intervention has many advantages (Krieger et al. 2005). However, some investigators have reported disappointing results (Thompson et al. 2008; McCauley et al. 2006), suggesting that much remains to be learned regarding community-based interventions.

Giving low-income families access to an effective vacuum cleaner removes a major source of environmental discrimination by increasing their ability to remove house dust (Krieger et al. 2005). Twenty-five percent of the families in the Seattle study did not have a vacuum cleaner at the beginning of the study; perhaps another 25% of homes did not have functioning vacuum cleaners, because they had full bags, worn or broken belts, air leaks, or worn agitators. The average surface dust loading, from the Seattle study, dropped from 2.64 to 0.96 g/m^2 after cleaning (Takaro et al. 2004). This compares favorably with the 1.4 g/m^2 median loading measured in 272 Midwest homes in the childhood leukemia study (Camann and Buckley 1994). The dust loading in both studies was measured with the HVS3. There is still room for improvement. A surface loading of 0.1 g/m^2 can be achieved when the removable deep dust is extracted (Roberts et al. 1999, 2004). Low-income families with asthmatic children, who live in substandard housing, may be exposed to hazards in addition to dust such as leaky roofs and plumbing, mold, pest intrusion, poor ventilation and temperature control, deteriorated paint, environmental tobacco smoke, poor access to medical care and food, crime, as well as stress related to complex social problems (Wu et al. 2007). The CHWs are trained to help such families deal with urgent difficulties that present barriers to reducing exposure to asthma triggers in the home.

5 Discussion

It is easy for families who receive home visits to monitor deep dust in carpets with the three-spot test. It may take 10–30 min of cleaning each day for a week to reduce dust-related risks by a factor of 10–100. Most families, with support of the local health department, health maintenance organization, and family doctor can take simple steps to reduce the exposure of infants to dust. Effective cleaning methods such as using an improved vacuum cleaner with higher efficiency dirt pickup and filtration, as well as a dirt finder, when combined with curtailing track-in of dirt may also save time. Even though there are many scientific uncertainties about the risks and health effects of dust exposure, it may be best to take the precautionary approach that emphasizes prevention and protection of children. It has also been argued that we have an ethical obligation to protect infants who have no political power and cannot protect themselves (Gilbert 2005).

Many pollutants get into house dust by migrating from consumer and household goods. Removing the most toxic compounds from consumer products is an important and necessary step in reducing exposure. However, it may be decades after a pollutant such as DDT, chlordane, or lead is banned, before it disappears from soil, food, indoor air, house dust, and carpets. A family can take action now, with improved cleaning and hand washing, to reduce a significant fraction of an infant's total exposure to allergens, lead, PBDEs, and other pollutants, by reducing contact with toxic compounds in house dust. This can provide an immediate reduction in exposure and enhances the large benefit of removing pollutants from consumer products, or emission and exposure reductions achieved by passage of the Clean Air Act of 1970 or other environmental laws that followed. These laws have reduced outdoor exposures and partially reduced indoor exposures and have reduced the rate at which pollutants build up in soil.

Additional cleaning is needed to protect babies with pica, because they tend to consume more dust. Effective cleaning and hand washing will also help protect people in several different situations: infants and toddlers living in old houses, where they may face risks of high-lead exposure; residents in areas where soil is contaminated by traffic (Beyea et al. 2006); those near Superfund sites; or near sources of other industrial emissions. It is estimated that 400,000 children (2% of children aged 5 or under) have ingested sufficient lead to cause learning disabilities or behavioral and gastrointestinal problems (CDC 2007; Markel 2007).

Replacing older carpets with the newer short-fiber impermeable ones appears to offer many advantages for reducing exposures. For many families, this is a more affordable option than installing bare floors. Some commercial carpets used in green hospitals come with low volatile organic emissions, are waterproof, 100% recyclable, easier to clean, and are free from PVCs (polyvinylchlorides), phthalates, and PBDEs, which are found in many older carpets (Walsh 2006).

Monitoring pollutant levels and setting standards for exposures have brought large benefits to workers. A similar effort would also benefit children. The cancer rates for US children increased by 26% from 1975 to 1998, with mortality rates falling by approximately 40%, because of improved treatments (U.S. EPA 2003), whereas the cancer mortality rate for US workers fell by 50% as a result of increased monitoring and control of carcinogen exposures (Trichopoulos et al. 1996). Asthma rates for children doubled since 1980 (Akinibami 2006), while occupational asthma from exposure to detergent dust was reduced as a result of monitoring and control efforts (Cathart et al. 1997). Monitoring and analysis of total exposure is necessary to efficiently manage health risks, disease, and costs associated with infants, workers, and adults (Ott 2006).

6 Research Recommendations

This review provides a basis for hypotheses that warrant future research:

1. The cost of home visits by CHWs or MHEs to reduce exposure to asthma triggers for children with poorly controlled asthma can be recovered in 1 yr or less in reduced medical costs.
2. The rates of childhood and adult asthma, and cancer, can be reduced by home interventions that reduce the total exposure of infants.
3. The percentage of children with chronic illness, including asthma and ADHD, can be reduced by home interventions that reduce the total exposure of infants to dust and other sources of pollutants.
4. Home visits will help families reduce dust exposures in the home by 90% in 1 wk, at a cost less than $160.
5. Home visits can be used to improve hand washing for families of infants.
6. Improved cleaning and hand washing can reduce the lead and PBDEs levels in infants, children, and adults.
7. Improved cleaning and hand washing can reduce the incidence of infections.

The National Children's Study (NCS) of 100,000 children over 21 yr (http://www.nationalchildrensstudy.gov/) offers an opportunity to test hypotheses 2, 3, 5, 6, and 7. A select group of the NCS children can serve as a control group for a matched intervention group.

7 Summary

The health risks to babies from pollutants in house dust may be 100 times greater than for adults. The young ingest more dust and are up to ten times more vulnerable to such exposures. House dust is the main exposure source for infants to allergens, lead, and PBDEs, as well as a major source of exposure to pesticides, PAHs, Gram-negative bacteria, arsenic, cadmium, chromium, phthalates, phenols, and other EDCs, mutagens, and carcinogens. Median or upper percentile concentrations in house dust of lead and several pesticides and PAHs may exceed health-based standards in North America.

Early contact with pollutants among the very young is associated with higher rates of chronic illness such as asthma, loss of intelligence, ADHD, and cancer in children and adults. The potential of infants, who live in areas with soil contaminated by automotive and industrial emissions, can be given more protection by improved home cleaning and hand washing. Babies who live in houses built before 1978 have a prospective need for protection against lead exposures; homes built before 1940 have even higher lead exposure risks. The concentration of pollutants in house dust may be 2–32 times higher than that found in the soil near a house.

Reducing infant exposures, at this critical time in their development, may reduce lifetime health costs, improve early learning, and increase adult productivity. Some interventions show a very rapid payback. Two large studies provide evidence that home visits to reduce the exposure of children with poorly controlled asthma triggers may return more than 100% on investment in 1 yr in reduced health costs. The tools provided to families during home visits, designed to reduce dust exposures, included vacuum cleaners with dirt finders and HEPA filtration, allergy control bedding covers, high-quality door mats, and HEPA air filters.

Infants receive their highest exposure to pollutants in dust at home, where they spend the most time, and where the family has the most mitigation control. Normal vacuum cleaning allows deep dust to build up in carpets where it can be brought to the surface and become airborne as a result of activity on the carpet. Vacuums with dirt finders allow families to use the three-spot test to monitor deep dust, which can reinforce good cleaning habits. Motivated families that receive home visits from trained outreach workers can monitor and reduce dust exposures by 90% or more in 1 wk. The cost of such visits is low considering the reduction of risks achieved. Improved home cleaning is one of the first results observed among families who receive home visits from MHEs and CHWs. We believe that proven intervention methods can reduce the exposure of infants to pollutants in house dust, while recognizing that much remains to be learned about improving the effectiveness of such methods.

Acknowledgments We thank Barbara Tombleson and Diane Dishion for their help in preparing and proofreading the manuscript as well as Bill Budd, Russell E. Crutcher, Greg Glass, Peter Hummer, Leah Michelson, and Michael G. Ruby for their contribution to house dust research presented in this manuscript. We also thank an anonymous reviewer for a very detailed review resulting in a much-improved manuscript.

References

Adgate JL, Willis RD, Buckley TJ, Chow JC, Watson JG, Rhoads GG, Lioy PJ (1998) Chemical mass balance source apportionment of lead in house dust. Environ Sci Technol 32:108–114.

Akinibami L (2006) The state of childhood asthma, United States, 1980–2005. Advance data from vital and health statistics: No 381. National Center for Health Statistics, Hyattsville, MD, pp. 1–24.

Arcury TA, Grzywacz, JG, Davis SW, Barr DB, Quandt SA (2006) Organophosphorous pesticide urinary metabolite levels of children in farmworker households in eastern North Carolina. Am J Ind Med 49:751–60.

Arlian LG, Morgan MS (2003) Biology, ecology, and prevalence of dust mites. Immunol Allergy Clin North Am 23:443–468.
ASTM (2007) Annual Book of Standards, Vol. 11.07. ASTM International, West Conshohoken, PA.
ATSDR (1993) Toxicological profile for 4,4-DDT, 4,4-DDE, and 4,4-DDD. U.S. Public Health Service, U.S. Department of Health and Human Services, Agency for Toxic Substances and Disease Registry, Atlanta, GA.
ATSDR (2000) Toxicological profile for polychlorinated biphenyls (PCBs). U.S. Department of Health and Human Services. Agency for Toxic Substances and Disease Registry, Atlanta, GA.
Benson AS (ed) (1985) Control of communicable disease in man. American Public Health Association, Washington, DC, p. 485.
Berube A (2007) Toward an economic analysis of the environmental burden of disease among Canadian children. J Toxicol Environ Health B 10:131–142.
Beyea J, Hatch M, Stellman SD, Santella RM, Teitelbaum SL, Prokopczyk B, Camann D, Gammon MD (2006) Validation and calibration of a model used to reconstruct historical exposure to polycyclic aromatic hydrocarbons for use in epidemiologic studies. Envrion Health Perspect 115:1053–1058.
Bornehag CG, Sundell J, Weschler CJ, Sigsgaard T, Lundgren B, Hasselgren M, Hagerhed-Engman L (2004) The association between asthma and allergenic symptoms in children and phthalates in house dust: A nested case-control study. Environ Health Perspect 112:1393–1397.
Bradman A, Whitaker D, Quiros L, Castorina R, Henn BC, Nishioka M, Morgan J, Barr DB, Harnly M, Brisbin JA, Sheldon LS, McKone TE, Eskenazi B (2007) Pesticides and their metabolites in the homes and urine of farmworker children living in the Salinas Valley, CA. J Expo Anal Environ Epidemiol 17:331–349.
Braun JA (2003) Home safe home: Preventing falls in the home through environmental assessment and modification. Geriatr Care Manag J. Summer Fall:8–12.
Braun JM, Kahn RS, Froelich T, Auinger P, Lanphear BP (2006) Exposures to toxicants and attention deficit hyperactivity disorder in U.S. children. Environ Health Perspect 114:1904–1909.
Brunekreef B, Dockery DW, Speixer FE, Ware JH, Spengler JD, Ferris BG (1989) Home dampness and respiratory morbidity in children. Am Rev Respir Dis 140:1363–1364.
Bunterngchit Y, Lockhart T, Woldstadt JC, Smith JL (2000) Age related effects of transitional floor surfaces and obstruction of view on gait related characteristics related to slips and falls. Int J Ind Ergon 25:223–232.
Butte W, Heinzow B (2002) Pollutants in house dust as indicators of indoor contamination. Rev Environ Contam Toxicol 175:1–46.
Calabrese EJ, Stanek E (1991) A guide to interpreting soil ingestion studies. II. Quantitative evidence of soil ingestion. Regul Toxicol Pharmacol 13:278–292.
Camann DE, Buckley JD (1994) Carpet Dust: An indicator of exposure at home to pesticides, PAHs, and tobacco, smoke. In: Proceedings of Sixth Conference of the International Society of Environmental Epidemiology and Fourth Conference of the International Society for Exposure Analysis: 18–21 September, Research Triangle Park, NC, and Princeton, NJ. International Society of Environmental Epidemiology. Paper No. 141.
Camann DE, Colt JS, Zuniga NM (2002a) Distribution and quality of pesticide, PAH, and PCB measurements in bag dust in four areas of U.S.A. In: Proceedings of Ninth International Conference on Indoor Air Quality and Climate: Indoor Air 2002, 30 June – 5 July, Monterey, CA, pp. 860–864.
Camann DE, Yau AY, Colt JS (2002b) Distributions and quality of acidic pesticide measurements in bag dust from four areas of USA. ISEA/ISEE Annual Conference, Vancouver, BC, Canada, Abstract 53.18, Aug. 2002.
Camann D, Yau A, Mc Pherson A (2005) Perfluorinated, organotins, PBDE, and other chemicals in house dust. In: Proceedings of the International Society of Exposure Analysis Conference, 30 Oct – 2 Nov, Tucson, AZ, and Princeton, NJ. International Society of Environmental Epidemiology.

Canfield RL, Henderswon CR, Cory-Slecta DA, Cox CR, Jusko TA, Lanphear BP (2003) Intellectual impairment in children with blood lead concentrations below 10 μg per deciliter. N Engl J Med 348:1517–1526.

CARB (2003) California portable classrooms study: Executive summary, Vol. 3. RTI International, California Air Resources Board, Sacramento, CA. Available at http://www.arb.ca.gov/research/indoor/pcs/pcs-fr/pcs_v3_es_03-23-04.pdf

Cathart M, Nicholson P, Roberts D, Bazley M, Juniper C, Murray P, Randall M (1997) Enzyme exposure, smoking and lung function in employees in the detergent industry over 20 years. Occup Med 47:473–478.

Centers for Disease Control and Prevention (CDC) (2007) CDC surveillance data 1997–2005. Centers for Disease Control and Prevention, Atlanta, GA.

Chuang JC, Callahan PJ, Lyu CW, Wilson NK (1999) Polycyclic aromatic hydrocarbon exposures of children in low-income families. J Expo Anal Environ Epidemiol 9:85–98.

Clickner RP, Marker D, Viet SM, Rogers J, Broene P (2001) National survey of lead and allergen in housing, Vol. 1: Analysis of lead hazards. Westat Inc./U.S. Dept. of Housing and Urban Renewal, Washington, DC.

Colt JS, Zahm SH, Camann DE, Hartge P (1998) Comparison of pesticides and other compounds in carpet dust samples collected from used vacuum cleaner bags and from the high volume surface sampler. Environ Health Perspect 106:721–724.

Colt JS, Lubin J, Camann DE, Davis S, Cerhan J, Severson RK, Bernstein L, Hartge P (2004) Comparison of pesticide levels in carpet dust and self reported pest treatment practices in four U.S. sites. J Expo Anal Environ Epidemiol 14:74–83.

Colt JS, Severson RK, Lubin J, Camann D, Davis S, Cerhan J, Cozen W, Rothman N, Hartge P (2005) Organochlorine compounds in carpet dust and risk of non-Hodgkin lymphoma. Epidemiology 16:516–525.

Colt JS, Davis S, Severson RK, Lynch CF, Cozen W, Camann D, Engels EA, Blair A, Hartge P (2006) Residential insecticide use and risk of non-Hodgkin's lymphoma. Cancer Epidemiol Biomarkers Prev 15:251–257.

Colt JS, Gunier RB, Metayer C, Nishioka MG, Bell EM, Reynolds P, Buffler PA, Ward MH (2008) Household vacuum cleaners vs. the high-volume surface sampler for collection of carpet dust samples in epidemiologic studies of children. BioMed Central. Environmental Health 7(6) (page numbers not available). Accessed Sept. 10, 2008 at http://www.ehjournal.net/content/7/1/6

Davies DJA, Thorton J, Watt JM, Culbard E, Harvey PG, Sherlock JC, Delves HT, Smart GA, Thomas FGA, Quinn MJ (1990) Lead intake and blood lead in two-year old U.K urban children. Sci Total Environ 90:13–29.

Dickey P (2005) Safer cleaning products. In: Dickey P (ed) Safer alternatives. Washington Toxics Coalition, Seattle, WA, pp. 5–8. http://www.watoxics.org/files/cleaning

Dickey P (ed) (2007) Master home environmentalist training manual. American Lung Association of Washington, Seattle, WA.

Dilley JA, Pizacani BA, Macdonald SC, Bardin J (2005) The burden of asthma in Washington State. Washington State Dept. of Health, DOH Pub. No. 345-201, Olympia, WA.

Egeghy PP, Qackenboss JJ, Catlin S, Ryan B (2005) Determinants of temporal variability in NEXHAS-Maryland environmental concentrations, exposures, and biomarkers. J Expo Anal Environ Epidemiol 15:388–397.

Eyre H, Kahn R, Roberson RM, Doyle C, Hong Y, Gansler T, Glynn T, Smith RA, Taubert K, Thun MJ (2004) Preventing cancer, cardiovascular disease, and diabetes: A common agenda for the American Cancer Society, the American Diabetes Association, and the American Heart Association. Circulation 109:3244–3255.

Farfel MR, Lees PSJ, Lim B, Rohde CA (1994) Comparison of two cyclone-based devices for the evaluation of lead-containing residential dusts. Appl Occup Environ Hyg 9:212–217.

Fenske RA, Lu C, Barr D, Needham L. (2002) Children's exposure to chlorpyrifos and parathion in an agricultural community in central Washington State. Environ Health Perspect 110:549–553.

Fernandez MF, Olmos B, Granada A, Lopez-Espinosa MJ, Molina-Molina JM, Fernandez JM, Cruz M, Olea-Serrano F, Olea N (2007) Human exposure to endocrine-disrupting chemicals and prenatal risk factors for cryptorchidism and hypospadias: A nested case-control study. Environ Health Perspect 115 (Suppl 1):8–14.

Franke DL, Cole EC, Leese KE, Foarde KK, Berry MA (1997) Cleaning for improved air quality: An initial assessment of effectiveness. Indoor Air 7:41–54.

Freeman NCG, Shalat SL, Black K, Jimenez M, Donnelly KC, Calvin A, Ramirez J (2004). Seasonal pesticide use in a rural community on the US/Mexico Border. J Exp Sci Environ Epidemiol 14:473–478.

General Accounting Office (GAO) (1999) Indoor pollution: Status of federal research activities. GAO/RCED-99-254. General Accounting Office, Washington, DC.

Gereda JE, Leung DYM, Thatayatikom A, Streib JE, Price MR, Klinnert MD, Liu AH (2000) Relation between house dust endotoxin exposure, type 1 T-cell development, and allergen sensitization in infants at high-risk of asthma. Lancet 355:1680–1683.

Gilbert SG (2005) Ethical, legal, and social issues: Our children's future. Neurotoxicology 26:521–530.

Grandjean P, Landrigan PJ (2006) Developmental neurotoxicity of industrial chemicals. Lancet 368:2167–2178. http://www.ncbi.nih.gov/pubmed/17174709

Hageman KJ, Simonich SL, Campbell DH, Wilson GR, Landers DH (2006). Atmospheric deposition of current-use and historic-use pesticides in snow at National Parks in the Western United States. Environ Sci Technol 40:3174–3180.

Hartge P, Colt JS, Severson RK, Cerhan JR, Cozen W, Camann D, Zahm SH, Davis S. (2005). Residential herbicide use and risk of non-Hodgkin lymphoma. Cancer Epidemiol Biomarkers Prev 14:934–937.

Howard PH (1991) Handbook of environmental fate and exposure data for organic chemicals, Vol. 3: Pesticides. Lewis, Ann Arbor, MI.

IARC (1985) IARC Monogr Eval Carcinog Risks Hum. Polynuclear aromatic compounds, bituminous, coal-tar and derived products, shale oils and soots, Vol. 35, Part 4. International Agency for Research on Cancer, Lyon, France.

IARC (2008) Overall evaluations of carcinogenicity to humans: Group 1: Carcinogenic to humans, Vol. 1-99. International Agency for Research on Cancer, Lyon, France. http://monographs.iarc.fr/ENG/Classification/crthgr01.php

Institute of Medicine (2000) Clearing the air: Asthma and indoor air exposures. National Academy Press, Washington, DC.

Jaakkola JJK, Hwang BF, Jaakkola N (2005) Home dampness, parental atophy, and asthma in childhood: A six-year population-based cohort study. Environ Health Perspect 113:357–361.

Jaakkola JJK, Ieromnimon A, Jaakkola MS (2006) Interior surface materials and asthma in adults: A population-based incident case-control study. Am J Epidemiol 164:742–749.

Jobling S, Tyler CR (2006) Introduction: The ecological relevance of chemically induced endocrine disruption in wildlife. Environ Health Perspect 114 (Suppl 1):7–8.

Jones KC, Strafford JA, Waterhouse KS, Furlong ET, Giger W, Hites RA, Schaffner C, Johnson AE (1989) Increases in the polynuclear aromatic hydrocarbon content of an agricultural soil over the last century. Environ Science Technol 23:95–101.

Jones-Otazo HA, Clark JP, Diamond ML, Archbold JA, Ferguson G, Harner T, Richardson GM, Ryan JJ, Wilford B (2005) Is house dust the missing exposure path for PBDEs? An analysis of urban fate and human exposure to PBDEs. Environ Sci Technol 38:5121–5130.

Juniper EF, Guyatt GH, Feeny DH (1996) Measuring quality of life in the parents of children with asthma. Quality Life Res 5:27–34.

Kattan M, Stearns S, Crain E, Stout J, Gergen P, Evans R, Visness CM, Gruchalla RS, Morgan WJ, O'Connor GT, Mastin JP, Mitchell HE (2005) Cost-effectiveness of a home-based environmental intervention for inner-city children with asthma. J Allergy Clin Immunol 116:1058–1063.

Kercsmar CM, Dearborn DG, Schluchter M, Xue L, Kirchner HL, Sobolewski J, Greenberg SJ, Vesper SJ, Allan T (2006) Reduction in asthma morbidity in children as a result of home remediation aimed at moisture sources. Environ Health Perspect 114:1574–1580.

Kim N, Ferguson J (1993) Concentrations of cadmium, copper, lead, and zinc in house dust in Christchurch, New Zealand. Sci Total Environ 138:1–21.

Kolarik B, Naydenov K, Larsson M, Bornehag CG, Sundell J (2008) The association between phthalates in dust and allergic diseases among Bulgarian children. Envron Health Perpect 116:98–103.

Krieger J, Takaro TK, Song L, Weaver M (2005) The Seattle-King County healthy homes project: A randomized, controlled trial of a community health worker intervention to decrease exposure to indoor asthma triggers. Am J Public Health 95:652–659.

Kurtz DA (1990) Long range transport of pesticides. Lewis, Ann Arbor MI.

Lanphear BP (1996) Lead-contaminated house dust and urban children's blood lead. Am J Public Health 86:1416–1421.

Lanphear BP, Esmond M, Jacobs DE, Weitzman M, Tanner M, Winter NL, Yaki B, Eberly S (1995). A side-by-side comparison of dust collection methods for sampling lead-contaminated house dust. Environ Res 68:114–123.

Lanphear BP, Hornung R, Khoury J, Yolton K, Baghurst P, Bellinger D, Canfield RL, Dietrich KN, Bornschein R, Greene T, Rothenberg SJ, Needleman HL, Schnaas L, Wasserman G, Graziano J, Roberts R (2005) Low-level environmental lead exposure and children's intellectual function: An international pooled analysis. Environ Health Perspect 113:894–899.

Larson LL, Lin SX, Gomez-Pichardo C, Della-Latta P (2004) Effect of antibacterial home cleaning and handwashing products on infectious disease symptoms: A randomized, double-blind trial. Ann Int Med 140:321–329.

Leung R, Koenig JQ, Simcox N, van Belle G, Fenske R, Gilbert SA (1997) Behavioral changes following participation in a home health promotion program in King County, Washington. Environ Health Perspect 105:1132–1135.

Lewis RD (2002) The Removal of lead contaminated house dust from carpets and upholstery. U.S. Department of Housing and Urban Renewal, Office of Healthy Homes and Lead Hazard Control, Washington, DC.

Lewis RD, Condoor S, Batek J, Ong KH, Backer D, Sterling D, Siria J, Chen JJ, Ashley P (2006) Removal of lead contaminated dusts from hard surfaces. Environ Sci Technol 40:590–594.

Lewis RG (2005) Residential post-application exposure monitoring. In: Franklin CA, Worgan JP (eds) Occupational and incidental residential exposure assessment. Wiley, Sussex, pp. 71–128.

Lewis RG, Lee RE Jr. (1976) Air pollution from pesticides: Sources, occurrences, and dispersion. In: Lee RE Jr. (ed) Air pollution from pesticides and agricultural processes. CRC Press, Boca Raton, FL, pp. 5-50.

Lewis RG, Fortmann RC, Camann DE (1994) Evaluation of methods for the monitoring of the potential exposure of small children to pesticides in the residential environment. Arch Environ Contam Toxicol 26:37–46.

Lewis RG, Fortune CR, Willis RD, Camann DE, Antley JT (1999) Distribution of pesticides and polycyclic aromatic hydrocarbons in house dust as a function of particle size. Environ Health Perspect 107:721–726.

Lewis RG, Fortune CR, Blanchard FT, Camann DE (2001) Movement and deposition of two organophosphorus pesticides within residences after interior and exterior applications. J Air Waste Manag Assoc 51:339–351.

Lioy PJ (2006) Employing dynamical and chemical processes for contaminant mixtures outdoors to the indoor environment: The implications for total human exposure analysis and prevention. J Expo Sci Environ Epidemiol 16:207–224.

Lioy PJ, Wainman T, Weisel C (1993). A wipe sampler for the quantitative measurement of dust on smooth surfaces: Laboratory performance studies. J Exp Anal Environ Epidemiol 3:315–330.

Lorber M (2008) Exposure of Americans to polybrominated diphenyl ethers. J Expo Sci Environ Epidemiol 18:2–19.

Louis GB, Damstra T, Diaz-Barriga, Faustman E, Hass U, Kavlock R, Kimmel C, Kimmel G, Krishnan K, Luderer U, Sheldon L (2007) Principles for evaluating health risks in children associated with exposure to chemicals, Environmental Health Criteria 237. World Health Organization, United Nations, New York.

Luby SP, Agboatwalla M, Felkin DR, Painter J, Billhimer W, Altaf A, Hoekstra RM (2005) Effect of hand washing on child health: A randomized controlled trial. Lancet 366:225–233.

Mackay D, Shiu WY, Ma KC (1992) Illustrated handbook of physical-chemical properties and environmental fate for organic chemicals, Vol. 1. Lewis, Boca Raton, FL.

Maertens RM, Bailey J, White PA (2004) The mutagenic hazards of settled house dust: A review. Mutat Res 567:401–425.

Maertens RM, Gagne RW, Douglas GR, Zhu J, White PA (2008a) Mutagenic and carcinogenic hazards of settled house dust. II. *Salmonella* mutagenicity. Environ Sci Technol 42:1754–1760.

Maertens RM, Yang X, Zhu J, Gagne RW, Douglas GR, White PA (2008b) Mutagenic and carcinogenic hazards of settled house dust. I. Polycyclic aromatic hydrocarbon cancer risk and preschool exposure. Environ Sci Technol 42:1747–1753.

Mahaffey KR, Annest JL (1985) Association of blood level, hand to mouth activity, and pica among children [Abstract]. Fed Proc 44:752.

Maki BE, Fernie GR (1990) Impact attenuation of floor coverings in simulated falling accidents. Ergonomics 21:107–114.

Markel H (2007) Getting the lead out: The Rhode Island lead paint trials and their impact on children's health. J Am Med Assoc 297:2773–2775.

McCauley LA, Lasarev MR, Higgens G, Rothlein J, Muniz J, Ebbert C, Phillips J (2001) Work characteristics and pesticide exposures among migrant agricultural families: A community-based research approach. Environ Health Perspect 109:533–538.

McCauley LA, Travers R, Lasarev M, Muniz J, Nailon R (2006) Effectiveness of cleaning practices in removing pesticides from home environments. J Agromed 11:81–88.

McCaustland KA, Bond WW, Bradley DW, Ebert JW, Maynard JE (1982) Survival of hepatitis in feces after drying and storage for 1 month. J Clin Microbiol 16:957–958.

McConnell LL, LeNoir JS, Datta S, Seiber J (1998) Wet deposition of current-use pesticides in the Sierra Nevada mountain range. Environ Toxicol Chem 17:1908–1916.

Meeker JD, Calafat AM, Hauser R (2007) Di(2-ethylhexyl) phthalate metabolites may alter thyroid hormone levels in men. Environ Health Perspect 115:1029–1034.

Menzie AC, MacDonell MM, Mumtaz M (2007) A phased approach for assessing combined effects from multiple stressors. Environ Health Perspect 115:807–816.

Micallef A, Caldwell J, Colls J (1998) The influence of human activity on the vertical distribution of airborne particle concentration in confined environments: Preliminary results. Indoor Air 8:131–136.

Miller RL, Garfinkel R, Horton M, Camann D, Perera FP, Whyatt R, Kinney PL (2004) Polycyclic aromatic hydrocarbons, environmental tobacco smoke, and respiratory symptoms in an inner-city birth cohort. Chest 126:1071–1078.

Morgan WJ, Crain EF, Gruchalla RS, O'Connor GT, Kattan M, Evans R, Stout J, Malindzak G, Smartt E, Plaut M, Walter M, Vaughn B, Mitchell H (2004) Results of a home-based environmental intervention among urban children with asthma. N Engl J Med 351:1068–1080.

Moriwaki H, Takata Y, Arakaw R (2003) Concentrations of perfluorooctane sulfonate (PFOS) and perfluorooctanic acid (PFOA) in vacuum cleaner dust collected in Japanese homes. J Environ Monit 5:753–757.

Nenstiel RO, White GL, Aikens T (1997) Handwashing: A century of evidence ignored. Clin Rev 7:55–62.

Nishioka MG, Burkholder HM, Brinkman MC, Gordon SM, Lewis RG (1996) Measuring transport of lawn-applied herbicide acids from turf to home: Correlation of dislodgeable 2,4-D turf residues with carpet dust and carpet surface residues. Environ Sci Technol 30:3313–3320.

Nishioka MG, Burkholder HM, Brinkman MC, Lewis RG (1999) Distribution of 2,4-dichlorophenoxyacetic acid in floor dust throughout homes following homeowner and commercial lawn applications: Quantitative effects of children, pets, and shoes. Environ Sci Technol 33:1359–1365.

Nishioka MG, Lewis RG, Brinkman MC, Burkholder HM, Hines C (2001) Distribution of 2,4-D in air and on surfaces following lawn applications: Comparing exposure estimates for young children from various media. Environ Health Perspect 109:1185–1191.

Occupational Safety and Health Administration (OHSA) (1987) Occupational exposure to formaldehyde. Fed Regist 52(223):46168.
Ott WR (2006) Exposure analysis: A receptor-oriented science. In: Ott WR, Steinemann AC, Wallace LA (eds) Exposure analysis. CRC Press/Taylor and Francis, Boca Raton, FL, pp. 3–32.
Papadopoulos A (1998) Semi- and non-volatile compounds in house dust. EUR 17748 EN, Joint Research Center, European Commission, Ispra, Italy.
Perera FP, Rauh V, Whyatt RM, Tsai W-Y, Tang D, Diaz D, Hoepner L, Barr D, Tu YH, Camann D, Kinney P (2006) Effect of prenatal exposure to airborne polycyclic aromatic hydrocarbons on neurodevelopment in the first three years of life among inner-city children. Environ Health Perspect 114:1287–1292.
Perrin JM, Bloom SH, Gortmaker SL (2007). The increase of childhood chronic conditions in the United States. J Am Med Assoc 297:2755–2759.
Pope AM, Patterson R, Burge H (1993) Indoor allergens: Assessing and controlling health effects. Institute of Medicine, National Academy Press, Washington, DC.
Primomo J, Johnston S, Dibase F, Nodolf J, Noren L (2006) Evaluation of a community-based outreach worker program for children with asthma. Public Health Nurs 23:234–241.
Que Hee SS, Peace B, Clark CS, Boyle JR, Bornstein RL, Hammond PB (1985). Evolution of efficient methods to sample lead sources, such as house dust and hand dust, in the homes of children. Environ Res 38:77–95.
Rasmussen PE (2004) Can metal concentrations in indoor dust be predicted from soil geochemistry? Can J Anal Sci Spectrosc 49:166–174.
Rasmussen PE, Subramanian KS, Jessiman BJ (2001) A multi-element profile of house dust in relation to exterior dust and soils in the city of Ottawa, Canada. Sci Total Environ 267:125–140.
Riseborough RW (1990) Beyond long-range transport: A model of a global gas chromatographic system. In: Kurtz, DA (ed) Long range transport of pesticides. Lewis, Chelsea, MI, pp. 417–426.
Roberts JW (2007) Reducing health risks from dust in the home. In: Dickey P (ed) Master home environmentalist training manual. American Lung Association of Washington, Seattle, WA, Chapter 4, pp. 4-1–4-14.
Roberts JW, Dickey P (1995) Exposure of children to pollutants in house dust and indoor air. Rev Environ Contam Toxicol 143:59–78.
Roberts JW, Ott WR (2006) Exposure to pollutants from house dust. In: Ott WR, Steinemann AC, Wallace LA (eds) Exposure analysis. CRC Press/Taylor and Francis, Boca Raton, FL, pp. 319–345.
Roberts JW, Ruby MG, Warren GR (1987) Mutagenic activity in house dust. In: Sandu SS, DeMarini DM, Mass MJ, Moore MM, Mumfod JL (eds) Short-term bioassays in the analysis of complex mixtures. Plenum, New York, pp. 355–367.
Roberts JW, Budd WT, Ruby, MG, Bond AE, Lewis RG, Wiener RG, Wiener W, Camann D (1991a). Development and field testing of a high volume sampler for pesticides and toxics in dust. J Expos Anal Environ Epidemiol 1:143–155.
Roberts JW, Camann DE, Spittler TM (1991b) Reducing lead exposure from remodeling and soil track-in in older homes. In: Proceedings of the Annual Meeting of the Air and Waste Management, Paper No. 134.2, June 16–21, 1991, Vancouver, Canada. Air and Waste Management Association, Pittsburgh, PA, pp. 1–15.
Roberts JW, Budd WT, Ruby MG, Stamper JR, Camann DE, Sheldon LS, Lewis RG (1991c). A high volume small surface sampler (HVS3) for pesticides and other toxics in house dust. In: Proceedings of the 84th National Meeting of the Air & Waste Management Association, Paper No. 91-150.2, Vancouver, BC, Canada.
Roberts JW, Crutcher III ER, Crutcher IV ER, Glass G, Spittler T (1996) Quantitative analysis of road and carpet dust on shoes. In: Measurement of toxic and related air pollutants, May 7–9, 1996, Research Triangle Park, NC. Air and Waste Management Association, Pittsburgh, PA, pp. 829–835.
Roberts JW, Clifford WS, Glass G, Hummer PG (1999) Reducing dust, lead, bacteria, and fungi in carpets by vacuuming. Arch Environ Contam Toxicol 36:477–484.

Roberts JW, Glass G, Mickelson L (2004) A pilot study of the measurement and control of deep dust, surface dust, and lead in 10 old carpets using the 3-spot test while vacuuming. Arch Environ Contam Toxicol 48:16–23.

Rodes CE, Peters TM, Lawless PA, Wallace LA (1996) Aerosol sampling biases in personal exposure measurements. [Abstract]. Sixth Annual Meeting of the International Society of Exposure Analysis, New Orleans, LA.

Rudel RA, Camann DE, Spengler JD, Korn LR, Brody JG (2003) Phthalates, alkylphenols, polybrominated diphenyl ethers, and other endocrine-disrupting compounds in indoor air and dust. Environ Sci Technol 37:4543–4553.

Rudel RA, Seryak LM, Brody JG (2008) PCB-containing wood floor finish is a likely source of elevated PCBs in residents' blood, household air and dust: A case study of exposure. Environ Health 7:2.

Salam MS, Li YF, Langholtz B, Gilliland FD (2003) Early life risk factors for asthma: Findings from the children's health study. Environ Health Perspect 112(6):760–765.

Sandora TJ, Taveras, Chih MC, Resncik EA, Lee GM, Ross-Degnan D, Goldman (2005) A randomized, controlled trial of a multifaceted intervention including alcohol-based hand sanitizer and hygiene education to reduce illness transmission in the home. Pediatrics 116(3):587–594.

Sears MR, Greene JM, Willan AR, Wiecek EM, Taylor DR, Flannery EM, Cowan JO, Herbison GP, Silva PA, Poulton R (2003) A longitudinal, population-based study of childhood asthma followed to adulthood. N Engl J Med 349:1414–1422.

Selgrade MK, Lemanske RF, Gilmour MI, Neas LM, Ward MDW, Henneberger PK, Weissman DN, Hoppin JA, Dietert RR, Sly PD, Geller AM, Enright PL, Backus GS, Bromberg PS, Germolec DR, Yeatts KB (2006) Induction of asthma and the environment: What we know and need to know. Environ Health Perspect 114:615–619.

Sheldon L, Clayton A, Keever J, Perritt R, Whittaker D (1992) PTEAM: Monitoring phthalates and PAHs in indoor and outdoor air samples in Riverside. California Air Resource Board, Sacramento, CA.

Siefert B, Becker K, Dieter H, Krause C, Schultz C, Siewert M (2000) The German environmental survey 1990/1992 (GerES II): Reference concentrations of selected environmental pollutants in blood, urine, hair, house dust, drinking water and indoor air. J Expo Anal Environ Epidemiol 10:552–565.

Simcox NJ, Fenske RA, Wolz SA, Lee IC, Kalman DA (1995) Pesticides in household dust and soil. Environ Health Perspect 103:1126–1134.

Smucker SJ (2004) Region IX preliminary remediation goals (PRGs) table. Memorandum from the Technical Support Section, U.S. EPA, Region 9, San Francisco, CA.

Stahlhut RW, Wijngaarden EV, Dye TD, Cook S, Swan SH (2007) Concentrations of urinary phthalate metabolites are associated with increased waist circumference and insulin resistance in adult U.S. males. Environ Health Perspect 115:876–882.

Stapleton HM, Dodder MG, Offenberg JH, Schantz MM, Wise SA (2005) Polybrominated diphenyl ethers in house dust and clothes dryer lint. Environ Sci Technol 38:925–931.

Stapleton HM, Allen JG, Kelly SM, Konstantinov A, Klosterhaus S, Watkins D, McClean MD, Webster TF (2008). Alternate and new brominated flame retardants detected in U.S. house dust. Environ Sci Technol 42:6910–6916.

Sullivan SD, Weiss KB, Lynn H, Mitchell H, Gergen PJ, Evans R (2002) The cost-effectiveness of inner-city asthma intervention for children. J Allergy Clin Immunol 110:576–581.

Swan SH, Main KM, Liu F, Stewart SL, Kruse RL, Calafat AM, Mao CS, Redmond JB, Ternand CL, Sullivan S, Teague JL, and Study for Future Families Research Team (2005) Decrease in anogenital distance among male infants with prenatal phthalate exposure. Environ Health Perspect 113:1056–1061.

Takaro TK, Krieger J, Song L, Beaudet N (2004) Effect of environmental interventions to reduce exposure to asthma triggers in homes of low-income children in Seattle. J Expo Anal Environ Epidemiol 14 (Suppl 1):S133–S143.

Thatcher TL, Layton DW (1995) Deposition, resuspension, and penetration of particles within a residence. Atmos Environ 29:1487–1497.

Thayer KA, Meinick R, Burns K, Davis D, Huff J (2005) Fundamental flaws of hormesis for public health decisions. Environ Health Perspect 113:1271–1276.

Thompson B, Coronado GD, Vigoren EM, Griffith WC, Fenske RA, IKlessel JC, Sirai JH, Faustman EM (2008) Para Ninos Saludables: A community intervention trial to reduce organophosphate pesticide exposure in children of farmworkers. Environ Health Perspect 116:687–694.

Thorne PS, Kulhánková K, Yin M, Cohn R, Arbes SJ Jr, Zeldin DC (2005) Endotoxin exposure is a risk factor for asthma: The National Survey of Endotoxin in United States Housing. Am J Respir Crit Care Med 172:1371–1377.

Tomlin CDS (2000) The pesticide manual, 12th Ed. British Crop Protection Council, Farnham, UK.

Tonne CC, Whyatt RM, Camann DE, Perera FP, Kinney PL (2004) Predictors of personal polycyclic aromatic hydrocarbon exposures among pregnant minority women in New York City. Environ Health Perspect 112:754–759.

TOXNET (2007) Toxicology data network. National Library of Medicine, National Institutes of Health, Washington, DC. http://toxmap.nlm.nih.gov/toxmap/main/chemPage.jsp?chem=Indeno(1,2,3-cd)pyrene).

Trichopoulos D, Li FP, Hunter DJ (1996) What causes cancer? Sci Am 275:80–87.

U.S. EPA (1987) Unfinished business: A comparative assessment of environmental problems. Overview. EPA 400:02. U.S. EPA, Washington, DC.

U.S. EPA (1989) Development of a high volume surface sampler for pesticides in floor dust. EPA 600/4-88/036. U.S. EPA, Research Triangle Park, NC.

U.S. EPA (1990) Reducing risk: Setting priorities and strategies for environmental protection. EPAS SAB-EC-900-021. U.S. EPA, Washington, DC.USEPA (1995) Sampling house dust for lead: Basic concepts and literature review. EPA 600-R-95-007. U.S. EPA, Research Triangle Park, NC.

U.S. EPA (1996a) Exposure factors handbook, Report No. 600/P-95/002. U.S. EPA, Washington, DC.

U.S. EPA (1996b) Evaluation of dust samplers for bare floors and upholstery. EPA/600/R-96/001. U.S. EPA, National Exposure Research Laboratory, Research Triangle Park, NC.

U.S. EPA (1997) Summary and assessment of published information on determining lead exposures and mitigating lead hazards associated with dust and soil in residential carpets, furniture, and forced air ducts. EPA 747-S-97-001. U.S. EPA, Washington, DC.

U.S. EPA (2000) Analysis of aged in-home carpeting to determine the distribution of pesticide residues between dust, carpet, and pad compartments. EPA/600/R-00/030. U.S. EPA, Research Triangle Park, NC. http://www.epa.gov/ord/WebPubs/carpet/600r00030.pdf

U.S. EPA (2002) Child specific exposure factors handbook. Report No. 600-P-002B. U.S. EPA, Washington, DC.

U.S. EPA (2003) America's children and the environment: Measures of contaminants, body burdens, and illnesses, 2nd Ed. EPA 240-R-03-001. U.S. EPA, Washington, DC.

U.S. EPA (2004) Pesticide industry sales and usage, 2000 and 2001 market estimates. Report No. 733-R-02-001. U.S. EPA, Washington, DC. http://www.epa.gov/oppbead1/pestsales/

U.S. EPA (2007) Important exposure factors for children: An analysis of laboratory and observations data characterizing cumulative exposure to pesticides. March 2007. National Exposure Research Laboratory. Office of Research and Development Research, Triangle Park, NC. EPA 600/R-07/013. http://www.epa.gov/nerl/research/data/exposure-factors.pdf

Walsh B (2006) PVC elimination prominent in green hospitals. Healthy Building News, 11 Oct.

Ward M, Lubin J, Giglierano J, Colt JS, Woter C, Bekiroglu N, Camann D, Hartge P, Nuckols JR (2006) Proximity to crops and residential exposure to agricultural herbicides in Iowa. Environ Health Perspect 114:893–897.

Ward MH, Colt JS, Metayer C, Gunier RB, Lubin J, Nishioka MG, Reynolds P, Buffler PA (2008) Residential exposure to polychlorinated biphenyls and organochlorine pesticides and risk of childhood leukemia. Abstract 1090, ISEE/ISEA 2008 Joint Annual Conference, Pasadena, CA.

Watson WA, Litovitz TL, Rogers GC Jr, Klein-Schwartz W, Reid N, Youniss J, Flanagan A, Wruk KM (2005) Annual report of the American association of poison control center's toxic exposure surveillance system. Am J Emerg Med 23:589–666.

Wilson NK, Chuang JC, Morgan MK, Lordo RA, Sheldon LS (2007) An observational study of the potential exposures of preschool children to pentachlorophenol, Bisphenol-A, and nonylphenol at home and daycare. Environ Res 103:9–20.

Wu F, Takaro TK (2007) Childhood asthma and environmental interventions. Environ Health Perspect 115:971–975.

Wu N, Herrmann T, Paeke O, Tickner J, Hale R, Harvey E, La Guardia M, McClean MD, Webster TF (2007) Human exposure to PBDEs: Association of PBDE body burdens with food consumption and house dust concentrations. Environ Sci Technol 41:1585–1589.

Zeldin DC, Eggleston P, Chapman M, Piediomonte G, Renz H, Peden D (2006) How exposure to biologics influence the induction and incidence of asthma. Environ Health Perspect 114:620–626.

Pulmonary Toxicity and Environmental Contamination: Radicals, Electron Transfer, and Protection by Antioxidants

Peter Kovacic and Ratnasamy Somanathan

Contents

1 Introduction 41
2 Potential Mechanisms of Pulmonary Toxicity 43
    2.1 Electron Transfer 43
    2.2 Reactive Oxygen Species 43
    2.3 Reactive Nitrogen Species 45
    2.4 Oxidative Stress 45
3 A Survey of Pulmonary Toxicants 45
    3.1 Gases and Vapors 45
    3.2 Liquids 49
    3.3 Anesthetics and Therapeutic Agents 53
    3.4 Other Environmental Toxicants 54
    3.5 Other (Radiation) 58
4 Illness and Oxidative Stress 58
    4.1 Asthma 58
    4.2 Chronic Obstructive Pulmonary Disease 59
5 Antioxidant Benefits and Pulmonary Toxicity 59
6 Summary 59
References 60

1 Introduction

The pulmonary system is one of the main targets for toxicity. In the industrial age, there has been a large increase in atmospheric pollutants, including industrial products, particulates (asbestos and silica), cigarette smoke, ozone, nitrogen oxides, and substantial number of miscellaneous materials. In lung tissues, many adverse reactions result from exposure to these pollutants; some principal ones include asthma, chronic obstructive pulmonary disease (COPD), and cancer.

P. Kovacic (✉)
Department of Chemistry, San Diego State University, San Diego, CA, 92182-1030, USA
e-mail: pkovacic@sundown.sdsu.edu

R. Somanathan
Centro de Graduados e Investigación del Instituto Tecnológico de Tijuana, Apdo postal 1166, Tijuana, B.C. Mexico

The emphasis of this review is on three mechanisms by which many pulmonary toxicants, usually as derived metabolites, induce their effects: electron transfer (ET) (electron movement from one site to another), reactive oxygen species (ROS), and oxidative stress (OS), involving cellular insults. The preponderance of bioactive substances or their metabolites have chemical groups that we believe may play an important role in the physiological responses connected with induction of pulmonary toxicity. Such chemical functionalities include quinones (or their phenolic precursors), metal complexes (or complexors), aromatic nitro compounds (or reduced hydroxylamine and nitroso derivatives), and conjugated imines (or iminium species).

In vivo redox cycling of inhaled oxygen can give rise to OS through generation of ROS such as hydrogen peroxide, hydroperoxides, alkyl peroxides, and diverse radicals (hydroxyl, alkoxyl, hydroperoxyl, and superoxide). In some cases, ET results in interference with normal electrical effects, e.g., in mitochondrial respiration or neurochemistry.

Generally, active chemical entities that possess ET groups display reduction potentials (a measure of ease of electron uptake) in the physiologically responsive range that is more positive than -0.5 V. ET, ROS, and OS have been increasingly implicated in the mode of action of drugs and toxins. Among these are the following: anti-infective agents (Kovacic and Becvar 2000), anticancer drugs (Kovacic and Osuna 2000), carcinogens (Kovacic and Jacintho 2001a), reproductive toxins (Kovacic and Jacintho 2001b), nephrotoxins (Kovacic et al. 2002), hepatotoxins (Poli et al. 1989), cardiovascular toxins (Kovacic and Thurn 2005), nerve toxins (Kovacic and Somanathan 2005), mitochondrial toxins (Kovacic et al. 2005), abused drugs (Kovacic and Cooksy 2005), and various other categories, including human illnesses (Halliwell and Gutteridge 1999). The generated ROS can be beneficial, as in the case of therapeutics, or harmful, as with toxicants.

This review demonstrates that the ET–ROS–OS unifying theme, which has been successful in describing effects with other classes of toxins, can also apply to pulmonary toxicants. Toxic chemical groups constitute a wide variety of structurally diverse substances. Not all chemical interactions involving ET–ROS–OS are deleterious ones. Numerous reports exist on beneficial effects of antioxidants (AOs) or benefits of cell signaling, in which a recent review addresses the hypothesis of electron and radical participation (Kovacic and Pozos 2006). This review reveals that the unifying theme based on ET–ROS–OS serves as a common thread for the large majority of reported pulmonary toxins.

There is a plethora of experimental evidence supporting the OS unifying theme in affecting physiological activity, including generation of the common ROS, lipoperoxidation, degradation products of oxidation, depletion of AOs, effect of exogenous AOs, DNA oxidation in cell signaling and cleavage products, as well as electrochemical data. This comprehensive, unifying mechanism is in keeping with the frequent observations that many ET substances display a variety of activities, for example, multiple-drug properties, as well as toxic effects.

It should be emphasized that physiological activity is often complex and multifaceted. Our objective in this review does not encompass extensive treatment of chemical modes of action, such as enzyme inhibition and allosteric effects. Rather, the following sections present an overview of toxicity and the unifying theme as it pertains to well-known pulmonary toxicants.

2 Potential Mechanisms of Pulmonary Toxicity

2.1 Electron Transfer

Some elaboration on the fundamental biochemistry of ET functionalities may be useful. Redox cycling occurs between hydroquinone (**1**; Fig. 1) and *p*-benzoquinone (**2**), and between catechol and *o*-benzoquinone, with generation of superoxide via ET to oxygen. Semiquinones act as intermediates via electron uptake by quinones. Various amino acids can operate as electron donors. Superoxide serves as a precursor to a variety of other ROS. The quinones can either be endogenous or exogenous. With aromatic nitro compounds, the reduced nitroso- (**3**) and hydroxylamine- (**4**) metabolites can similarly enter into redox cycling, as can an oxy radical intermediate. Such compounds are in the exogenous group. Less well known are conjugated iminium (**5**) compounds, of which paraquat is a predominant member. Electron uptake yields resonance contributors (**6** and **7**). The imine precursor, commonly formed by condensation of protein primary amino groups with carbonyls, is readily converted to **5** by protonation.

2.2 Reactive Oxygen Species

ROS can arise from both endogenous and exogenous sources (Witschi 1997). Partial reduction of O_2 may occur to yield ROS, for example, superoxide. When cellular injury takes place release of chemical species, such as iron, into the extracellular space can lead to generation of deleterious ROS through the action of neutrophils and macrophages, which are adept at transforming oxygen into ROS. Neutrophils and macrophages are responsible for eliminating foreign organisms, and their actions in doing so may yield the undesirable effect of creating OS in normal cells. The lung is particularly susceptible to injury by oxygen. For example, hyperoxia damages endothelial and alveolar epithelial cells by generation of ROS (Halliwell and Gutteridge 1999). Exposure to oxygen results in increased intracellular generation of superoxide, and hence of other ROS, such as hydrogen peroxide, with ensuing effects from auto-oxidation reactions.

Adult respiratory distress syndrome (ARDS) is a dramatic example of lung toxicity involving ROS; it results from trauma, shock, sepsis, vomiting, and inhalation of toxins. About 50–90% mortality may result from this condition, which is characterized by pulmonary edema and alveolar damage. Inflammation, a common result of lung insult by toxic substances, is a precursor to events triggered by ROS.

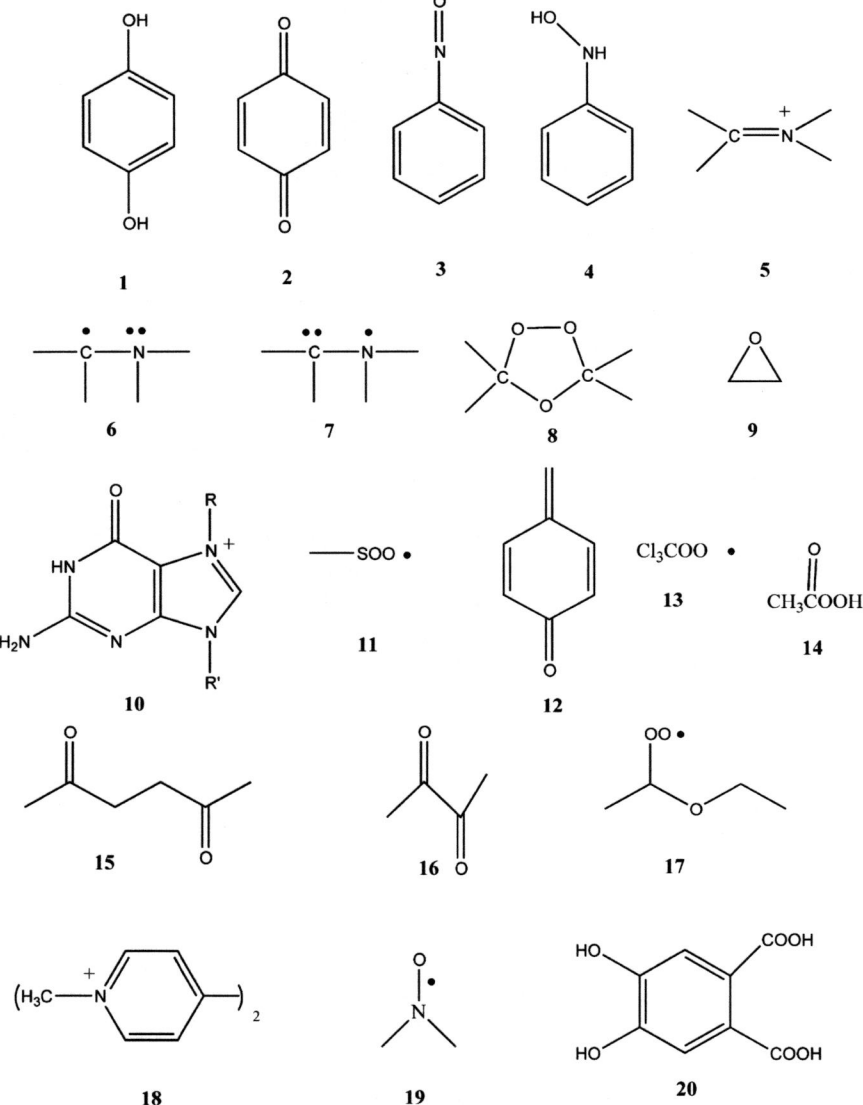

Fig. 1 Structure of pulmonary toxicants, metabolites, and reactive oxygen species in relation to mechanism

There is considerable evidence that OS is a contributing factor in ARDS (Halliwell and Gutteridge 1999). Witschi (1997) points out the following: "In general, free radicals represent an important component in the pathogenesis of lung disease." The role of ROS in lung damage is buttressed by the increased activity of free-radical-scavenging enzymes in lungs challenged by a variety of toxicants.

An appreciable literature on ROS and pulmonary toxicity exists and key citations from recent years include the following: Zaher et al. 2007; Lee et al. 2006; Wang et al. 2007; Manoury et al. 2005; Briede et al. 2004; Casalino-Matsuda et al. 2004; Yin et al. 2004; Woo et al. 2003; Langen et al. 2003; Kelly and Mudway 2003; Brown et al. 2004a; Wedgwood and Black 2003; Vassilakopoulos and Hussain 2007; Hammerschmidt et al. 2004; Dazy et al. 2003; Watt et al. 2004.

2.3 Reactive Nitrogen Species

Reports on this topic are less common than are those for ROS. Examples of RNS (reactive nitrogen species) include nitric oxide (Hickman-Davis et al. 2001; Bebok et al. 2002; van der Vliet et al. 1999; Machado et al. 2004; Puhakka 2005) and nitrogen oxides (Yuan et al. 2003; Qiu et al. 2004). A related participant is peroxynitrite derived by reaction of NO with superoxide. Hence, mechanistically, there is a relationship between RNS and ROS. The extensive general literature on NO is addressed in a book (Halliwell and Gutteridge 1999). Nitrogen dioxide is more toxic than NO, and on exposure it produces terminal bronchiolitis and lung edema (Henderson and Belinsky 1993).

2.4 Oxidative Stress

Oxidative stress results from an imbalance between pro-oxidants and AOs in favor of the former resulting in undesirable effects (Halliwell and Gutteridge 1999).

Although numerous articles deal with OS, some of those that are more cogent and recent include the following: Beeh et al. 2002; Rahman 2003; Bowler and Crapo 2002; Pinho et al., 2002; Cho and Kleeberger 2007; Ahamad et al. 2006; Barreiro et al. 2005. The contents of these articles address ROS, genetic mechanisms, cell signaling, AOs, and organ dysfunction.

3 A Survey of Pulmonary Toxicants

3.1 Gases and Vapors

3.1.1 Ozone

Ozone air pollution, a major environmental problem, arises mainly from the existence of engine exhaust gases and natural hydrocarbons (Witschi 1997). Pulmonary inflammatory lesions are the principal adverse effect from exposure to ozone, with prolonged exposure giving rise to various morphological and biochemical tissue changes. A number of metabolites, such as ozonides and hydroperoxides, formed from unsaturated lipids, and oxidation products of phenols could serve as radical precursors. Moreover, toxic entities, many of which are conceivably ROS, can arise from the interaction of ozone with other pollutants, for example, nitrogen oxides. AOs present in pulmonary tissues may provide protection against such toxicants. Ozone gas, a much more powerful oxidant than O_2, appears to exert its toxicity by direct oxidation, although radicals may also be involved (Halliwell and Gutteridge 1999; Miller et al. 1993). An important interaction between ozone and unsaturated fats produces ozonides (**8**), members of the ROS class; this class also includes hydrogen peroxide. Other topics that enjoin ozone and RNS include NO in cell signaling (Fakhrzadeh et al. 2004a, b), antioxidant responses to OS (Valacchi et al. 2004), ROS from macrophages and inflammatory mediators (Laskin et al. 2004), and oxidative damage to DNA (Ito et al. 2005).

3.1.2 Ethylene Oxide

Ethylene oxide is an industrial starting material primarily used in the manufacture of ethylene glycol and polyester. Inhaled ethylene oxide (**9**) gas induced lung neoplasms (Hong et al. 2007), in addition to various other toxic manifestations (Seguy et al. 1994). An important part of this agent's toxic action results from formation of adducts with DNA (Wu et al. 1999). Alkylation of DNA bases may occur with accompanying ROS formation (Kovacic and Jacintho 2001a). Reaction with guanosine can produce a conjugated iminium (**10**), which may operate as an ET agent. The mode of action responsible for DNA adduction and OS is applicable to other alkylators, such as sulfur mustard, sarin, PAH (polycyclic aromatic hydrocarbons), butadiene, styrene, TCE (trichloroethelyene), decadienal, and acrolein.

3.1.3 Sulfur Mustard

Sulfur mustard is used as a chemical warfare agent. It produces damage to the surfactant that coats the inner airways of the lung and also produces bronchioconstriction and inflammation of the respiratory tract that results in asthma-like symptoms (Van Helden et al. 2004). Sulfur mustard also causes hemorrhage and edema (Emmler et al. 2007) in pulmonary tissues. From a study of alterations in mouse lung involving GSH (glutathione) protein, DNA, lipid peroxidation, SOD (superoxide dismutase), and catalase, Elsayed and Omaye (2004) concluded that the bronchial changes from exposure to sulfur mustard are consistent with free-radical-mediated oxidative stress.

3.1.4 Sarin

The lung pathology produced by the warfare agent sarin has been presented in detail elsewhere (Pant et al. 1993). Mechanistic aspects of sarin's action, including enzyme inhibition and ET–OS, are covered in another review (Kovacic 2003).

Sarin, an organophosphate poison, disrupts normal function by operating as an alkylator of DNA. Alkylating agents are well-known generators of ROS, but the mechanisms by which this happens are not well elucidated.

3.1.5 Formaldehyde

Formaldehyde is an industrial chemical widely used in industry to manufacture building products, other chemicals, and household products. Exposure to formaldehyde as an air pollutant induces DNA damage (strand breaks and cross-links) in the lung (Xi et al. 2004). When such damage occurs, oxidative routes that result in genotoxicity may be involved. Yu et al. (2004) reported on formaldehyde-induced effects, including free radical injury and lipid peroxidation in the lung, and a decrease in AO status from inhibition of SOD and GSH peroxidase (Zhang et al. 2004). Kovacic and Somanathan (2007) reviewed formaldehyde-induced lung carcinogenicity that involved formation of ROS.

3.1.6 Phosgene

Phosgene is a valued industrial reagent and building block in organic synthesis. An investigation of pulmonary edema by phosgene revealed oxidative injury and effects on AO levels (Liang and Wang 2004; Sciuto et al. 2003). The reactive phosgene acid chloride can cause damage to a variety of cellular constituents, such as amines and alcohols. Such attacks on enzymes would have serious physiological consequences.

3.1.7 Carbon Monoxide

Carbon monoxide (CO) causes many accidental deaths and is often used as an agent in suicides. The affinity of CO for hemoglobin, to form carboxyhemoglobin, is the principal cause of mortality, although other cellular targets exist (Ryter and Otterbein 2004). Once combined with CO, hemoglobin cannot link with oxygen, thereby causing hypoxia. CO also affects several intracellular signaling pathways including guanylate cyclase and protein kinases (MAPK) that mediate vasoregulatory, antiinflammatory, antiapoptotic, and antiproliferative effects. Metabolism of CO that may result in neurotoxicity is addressed elsewhere, and it involves inactivation of essential metal enzymes by complexation (Kovacic and Somanathan 2005).

3.1.8 Hydrogen Cyanide

Battacharya et al. (1994) showed that intake of hydrogen cyanide gas induced various adverse influences including a direct effect on pulmonary cells, as evidenced by decreased compliance. The cyanide ion, a strong nucleophile, is probably the actual poison and acts by inhibiting essential Fe-containing constituents, such as hemoglobin and enzymes through strong ligation.

3.1.9 Hydrogen Sulfide

About 10% of total global emissions of hydrogen sulfide come from petroleum refineries. Baskar et al. (2007) reported that this gas exhibits various toxic effects to the pulmonary system including DNA damage, cytochrome oxidase inhibition (Dorman et al. 2002), hemorrhagic edema (Green et al. 1991) and it accentuates respiratory symptoms (Hessel et al. 1997). Several studies implicate hydrogen sulfide in inducing OS during tissue insults; such OS results from radical metabolites (Attene-Ramos et al. 2007; Pryor et al. 2006). These radicals are of type **11**. In a recent review, Kovacic and Somanathan (2006a) reported that vitamins C and E and thiols, which are usually classified as AOs, can also produce ROS under appropriate conditions.

3.1.10 Sulfur Dioxide

Sulfur dioxide is a common atmospheric pollutant. People are also exposed to it in some workplace settings. It is produced when sulfur-containing fuels such as coal and oil are burned, during smelting of metals and by certain other industrial processes. Xie and Fan (2007) reported sulfur dioxide-induced protein oxidation and DNA-protein cross-linking in the lung. Meng and Liu (2007) showed that cell morphological changes are partly the result of OS caused by sulfur dioxide. More recently, Ergonul et al. (2007) showed that the antioxidant vitamin E protects against sulfur dioxide-induced lipid peroxidation. Oxidative damage from sulfur dioxide exposure appears to be associated with a decrease in AO activity of various AO systems (Meng et al. 2003). Radical anions of SO_3, SO_4, and SO_5 formed during sulfite oxidation also produce toxic effects and are discussed elsewhere (Kovacic and Somanathan 2005).

3.1.11 Chlorine and Hypochlorous Acid

Chlorine continues to be widely used as a bleaching agent, disinfectant, and oxidizer. As result, it is present in many industrial and municipal wastewaters in concentrations ranging from a few ppb to 1% or more. Grasemann et al. (2007) reported that exposure to chlorine gas causes occupational asthma. Observed effects include airway inflammation and NO formation. The toxic effects that result from NO exposure have been documented in a previous review by Jacintho and Kovacic (2003). *In vivo*, chlorine readily reacts with water to generate the oxidant hypochlorous acid, which is also toxic. Behavior of chlorine as a ROS has also been addressed in the literature (Halliwell and Gutteridge 1999).

3.2 Liquids

3.2.1 Benzene

Yoon et al. (2004) studied the pulmonary pathogenesis from exposure to benzene. Their work revealed pneumotoxic and mucin-induced secretions as effects from benzene exposure. Studies on benzene-induced carcinogenicity indicate that production of ROS occurs, apparently by redox cycling of quinone metabolites (Kovacic and Jacintho 2001a).

3.2.2 Toluene

In a study on lung microsomes, Pyykkö (1983) showed that activities of monooxygenases and concentration of cytochrome P-450 decreased after toluene was inhaled. Lung exposure to toluene results in oxidative damage to lungs, effects on DNA and may result from formation of 8-hydroxy-D-guanosine (8-OH-DG; Tokunaga et al. 2003). The origin of the ROS involved entails the hydroxylated metabolites, quinone methide (**12**), and epoxide (Kovacic and Somanathan 2005).

3.2.3 Styrene

Styrene is an important starting material in the polymer industry. The monomer generates superoxide and TBARS (thiobarbituric acid reactive species), both indicative of OS (Vettori et al. 2005). There is marked oxidative DNA damage and evidence for induction of apoptosis. 2-Phenyloxirane is the main styrene metabolite. This epoxide is susceptible to nucleophilic attack from the amine groups of nucleic acid bases and may produce DNA adducts; it also depletes GSH (Csanady et al. 2003). The epoxide was also examined in a toxicokinetics study. Kaufmann et al. (2005) noted that styrene causes lung tumors in mice. Cruzan et al. (2005) reported that the epoxide and 4-vinylphenol metabolites caused GSH depletion and changes in the bronchiolar epithelium. Ring-oxidized metabolites are more potent than the epoxide in inducing pneumotoxicity. A similar mechanism involving a reactive epoxide is envisaged for other alkenes, such as TCE, acrolein, butadiene, and decadienal.

3.2.4 Trichloroethylene

Chen et al. (2002) showed that GSH plays an important role in protecting against TCE-induced oxidative stress and apoptosis in human lung cancer cells; such protection may be mediated through cell signaling. In another study (Giovanetti et al. 1998), evidence of OS was gleaned from observed decreases in GSH. There is also evidence for the formation of an epoxide metabolite that may also induce OS (Kovacic et al. 2002).

3.2.5 Acrolein

Acrolein is a widely distributed environmental pollutant. Moreover, the reactive alpha, beta unsaturated aldehyde is present in relatively high concentrations in cigarette smoke. A study was made of DNA adducts in the human lung (Zhang et al. 2007). The reaction between acrolein and certain DNA constituents could be initiated via the reactive aldehyde group, followed by ROS interaction with the double bond leading to an epoxide, which is susceptible to further nucleophilic attack. The reaction products produce lipid peroxidation and GSH depletion (Halliwell and Gutteridge 1999). Protection of lung tissues from the effects of acrolein is provided by thiols and Fe chelators.

3.2.6 Epichlorohydrin

Luo et al. (2004) showed that this epoxide produces obstructive lung abnormalities and airway damage, and such damage may be affected by titers of GSH S-transferase. Epichlorohydrin, as an industrial chemical, is a weak–moderate oncogen that binds to macromolecules *in vivo* (Mazzullo et al. 1984). Epichlorohydrin can function as a powerful alkylator of DNA; the reaction with DNA is initiated by nucleophilic attack of an amine group present in the DNA bases on the epoxide carbon or via the carbon bearing the halide group leading to a variety of harmful intermediates (Kovacic and Jacintho 2001b).

3.2.7 Chloroform

Chloroform is a useful industrial solvent with applications in the manufacture of plastics, refrigerants, and as an extraction solvent for fats, oils, waxes, resins, gums, and adhesives. In an earlier work, Torkelson et al. (1976) showed that chloroform vapor caused adverse effects in a variety of animals, including rats, rabbits, dogs, and guinea pigs. Metabolism of chloroform yields the trichloromethyl radical that forms the peroxyl radical **13** on reaction with oxygen; this reaction product can then serve as precursor to other ROS (Kovacic et al. 2002). A similar mechanism is also envisaged to occur with other harmful alkyl halides.

3.2.8 Carbon Tetrachloride

Carbon tetrachloride, a man-made chemical, is used as a solvent and in the manufacture of other chemicals. Carbon tetrachloride gas is an ozone-depleting substance and is banned from many major chemical manufacturing processes. Exposure to carbon tetrachloride gas induced changes in mixed function oxidases and cytochromes (Chen et al. 1977). On exposure, lipid peroxidation appears to occur in lung microsomes (Willis and Recknagal 1979) and may result in fibrosis

(Pääkkö et al. 1996). Metabolism of carbon tetrachloride leads to ROS by a process described elsewhere (Kovacic and Jacintho 2001a); the process includes participation of ROS such as **13**.

3.2.9 Ethanol

Laryngeal cancer risk is associated with alcohol and tobacco consumption (Risch et al. 2003). In rat lungs, ethanol exposure results in an increase in ROS and lipid peroxidation (Jurczyk et al. 2003). Chronic ethanol ingestion enhanced lung injury and decreased GSH concentrations (Brown et al. 2004b). A 2006 review details various metabolic routes for ethanol that generate ROS, and the associated toxic acetaldehyde, which can form **14** (Kovacic and Somanathan 2006b).

3.2.10 *n*-Hexane

Yang et al. (2006) and Sahu et al. (1982) showed that the inhaled vapors of *n*-hexane caused cell damage accompanied by enhanced pulmonary secretions. Short-term exposure affected the amount and composition of pulmonary surfactant (Hadjiivanova et al. 1987). Long-term contact revealed morphological changes consisting of air space enlargement, fibrosis, and papillary tumors (Lungarella et al. 1984). Metabolic data revealed that conversion to 2,5-hexanedione (**15**) occurred, which condenses with protein primary amine to yield the toxic product (Kovacic and Somanathan 2005).

3.2.11 Paint Thinner

Inhalation of the vapors of paint thinner solvent mixtures by young people constitutes a serious societal health problem (IIgazli et al. 2004); industrial exposure also occurs. Intake in high doses causes harm to the respiratory tract, partly through generation of OS. Toluene and methanol are major paint thinner constituents, accompanied by acetone, acetals, alcohols, turpentine, mineral spirits, and other solvents.

3.2.12 Carbon Disulfide

Carbon disulfide is a hazardous side product generated when sodium cellulose xanthogenate is precipitated in acid baths in the cellulose manufacturing industry. On exposure, disturbances in metabolic processes of the lung were observed, including enzymatic activity and protein synthesis (Mikhaïlova et al. 1987). Toxic effects, for example, edema and neoplasia, appear to result, in part, from metabolism to reactive intermediates, such as sulfur oxides and sulfur dioxides, which bind to macromolecules (Neal and Halpert 1982). Radical products have been reported, including ROS, such as **11** (Kovacic and Somanathan 2005). Radical cations were detected that interacted with DNA.

3.2.13 Diacetyl

Diacetyl is a natural byproduct of fermentation and is used as a flavoring agent that gives popcorn its butter-like taste. Bronchilitis obliterans, a serious lung disease, arises in workers exposed to butter flavoring volatiles (particularly diacetyl) in microwave popcorn plants (Fedan et al. 2006). Inhalation of diacetyl compromises epithelial barrier function. Kovacic (2007) addressed the mode of action of diacetyl at the molecular level, which involves ET processes arising from the conjugated α-dicarbonyl (**16**).

3.2.14 Phenols

Pentachlorophenol, a toxic pesticide, causes chronic lung damage when ingested and inhaled (Proudfoot 2003). The relationship of the ET-ROS-OS framework to metabolism of phenols, in which an ET quinone plays a key role, is treated elsewhere (Kovacic and Jacintho 2001b). 2,6-Di-*tert*-butyl-4-methylphenol (BHT) produces severe pulmonary inflammation and appears to be a major contributor to lung tumor promotion (Meier et al. 2007). Quinone methide metabolites, related to **13**, inhibit AO enzymes, leading to enhanced levels of ROS and inflammation. The ROS may also play a role in tumor promotion.

3.2.15 1,3-Butadiene

This industrial chemical, a multisite carcinogen, is metabolized to several epoxides that form adducts with DNA and proteins (Boysen et al. 2004). In one investigation this chemical was found to cause genetic damage, mutations, and cell cycle perturbation (Schmiederer et al. 2005). The diepoxide of 1,3-butadiene appears to be the most active metabolite and can produce harmful adducts by undergoing nucleophilic reactions with amines in nucleic acids and proteins.

3.2.16 2,4-Decadienal

Chang et al. (2005) reported that this unsaturated aldehyde, a product of lipid peroxidation, is associated with lung adenocarcinoma arising from exposure to cooking oil fumes. Results demonstrate an increase in OS, enhanced ROS, and decrease in GSH. Cotreatment with *N*-acetylcysteine prevented cell proliferation and release of cytokines. The model compound acrolein, structurally similar to 2,4-decadienal (also called heptenyl acrolein), serves as a source of ROS-OS in vivo (Kovacic 2006).

3.3 Anesthetics and Therapeutic Agents

3.3.1 Anesthetics

Inhaled anesthetics, such as chloroform, fluroxene, methoxyflurane, flurane, and enflurane, reversibly inhibit mitochondrial electron transport, resulting in interference with brain energy production and utilization (Levitt 1975). The toxicity of these agents, we believe, is related to the character of their intermediary or final metabolites. Chloroform, which has been supplanted as an anesthetic, forms radical metabolites as previously explained. Ethyl ether, also used in earlier times, can form a carbon radical adjacent to oxygen, followed by generation of the peroxyl radical (**17**) with subsequent formation of other ROS (Kovacic and Jacintho 2001b). More recent anesthetics, such as halothane, incorporate various halogens, which can be metabolized to radicals, related to **17**. Such radicals may include ROS that cause cell damage via lipid peroxidation (Durak et al. 1996; Tomasi et al. 1983).

3.3.2 Therapeutic Agents

General

Beltron and Lee (2001) address various respiratory disorders arising from therapeutic drugs. Included among the drugs are β-lactams, aspirin, mitomycin, methotrexate, methyldopa, hydralazine, Hg salts, nitrofurantoin, penicillamine, amphotericin, paclitaxel, and radiocontrast agents. In recent reviews the relationship between the parent compounds or, more commonly, their metabolites and the ET–ROS–OS framework is summarized (Kovacic and Becvar 2000; Kovacic and Osuna 2000).

Bleomycin

Bleomycin, a commonly used agent in cancer chemotherapy, can precipitate interstitial pulmonary fibrosis with a mortality of 1–2% (Witschi 1997). Mechanistically, the toxin binds to Fe(II), followed by conversion to an oxygen-ligated ferric-bleomycin complex, which may be a hydroperoxide (Kovacic and Jacintho 2003). Witschi (1997) noted that the subsequent attack on DNA results in chain cleavage via free radical involvement. Exposure to hyperoxia greatly potentiates lung injury; this portends a role for O_2. An efficient curative treatment entails a protocol that focuses on the second stage of toxicity, namely, fibroblast proliferation and collagen deposition. Administration of the antioxidant *N*-acetylcysteine via aerosol attenuates lung fibrosis induced by the drug in mice (Hagiwara et al. 2000).

3.4 Other Environmental Toxicants

3.4.1 Metals and Metal Compounds

This class is known to produce toxicity in a variety of organs (Kovacic and Somanathan 2005). With regard to electrochemistry, the reduction potentials of heavier metals are generally quite amenable to ET in biosystems. Thus, in the presence of some metals, for example, Fe, Cr, Ni, As, and Pb, redox cycling may result in formation of toxic ROS. The size and positive charge of metals are factors that affect electron uptake. In one report (Kovacic and Pozos 2006), metals are seen as being involved in cell signaling from the perspective of ET–ROS. Moreover, Valavanidis et al. (2005) revealed experiments in which the presence of redox-active transition metals in particulate matter (via Fenton reactions) appears to be involved in formation of hydroxyl radicals. Similar results entailing ROS were observed with particulates (Lewis et al. 2003; Salnikow et al. 2000). OS was generated by inhalation of the following metal species: Mn (Erikson et al. 2004; Dobson et al. 2003), Fe (Turi et al. 2004), Cr (Ortega et al. 2005), and As (Shi et al. 2004). GSH levels decreased in the olfactory bulb of male rats exposed to Mn, a response that is indicative of OS (Erikson et al. 2004). ROS, gene activation, and cell signaling pathways are discussed in relation to lung diseases induced by Cr and Ni (Barchowsky and O'Hara 2003). Results indicate that multiple oxidative species are involved in vanadium-induced lung inflammation and apoptosis (Wang et al. 2003). Prows et al. (2003) performed work that identified genetic susceptibility to Ni-induced lung injury.

3.4.2 Particulates

A 2007 book covers various aspects of particle toxicology, including types of damage, cell signaling, genotoxic effects, oxidative and nitrosative stress, and involvement of the immune system (Donaldson and Borm 2007). *In vitro* and in vivo animal studies demonstrate various responses to particulate exposure of the lung, such as increased cytokine, chemokine, and intracellular ROS generation, pulmonary inflammation, and development of alveolar epithelial hyperplasia and pulmonary fibrosis (Waldman et al. 2007). Involvement of oxidative stress is a common topic of such studies (Gonzales-Flecha 2004; Beck-Speier et al. 2005). ROS generation may produce DNA damage (Lai et al. 2005), genotoxicity, and mutagenesis (Knaapen et al. 2004). Donaldson et al (2004) overviewed induction of OS-responsive signaling pathways. Rhoden et al. (2004) found that lung inflammation is prevented by the AO *N*-acetylcysteine. More detailed discussion of ROS involvement from particulates can be found in the sections of this article that address asbestos, silica, and PAHs.

9,10-Phenanthroquinone, a major quinone contained in atmospheric particulates, generates ROS through redox cycling that causes iron-mediated oxidative damage (Sugimoto et al. 2005). Analysis showed that the quinone-enriched fraction from particulates is more potent than the PAH fraction as regards generation of ROS and induction of apoptosis (Xia et al. 2004).

3.4.3 Asbestos

Deposition of fibers in the lung over extended periods, apparently results in continual release of ROS. Concomitantly, iron mobilized from asbestos may contribute to OS via the Fenton reaction (Mossman and Gee, 1993). In cell-free systems, asbestos fiber can produce superoxide and hydroxyl radicals in the presence of H_2O_2. The bulk of evidence supports the contention that OS represents a fundamental mechanism of this toxicity. In tracheal cell cultures, scavengers of ROS inhibited asbestos-induced insults.

A review by Chen and Vallyathan (2005) identified various mechanisms by which fibers induce carcinogenesis. Induction of DNA damage and apoptosis apparently involve ROS and RNS mechanisms in altering mitochondrial function and activating death receptor pathways in cells (Upadhyay and Kamp 2003). Oxidants play important roles in initiation of numerous signal transduction pathways that are linked to apoptosis (Shukla et al. 2003a, b).

3.4.4 Silica

Exposure to particulate silica produces inflammation accompanied by generation of oxidants in the alveolar space (Fubini and Hubbard 2003). The ROS are formed not only on the particle surface, but also by phagocytic cells attempting to destroy the foreign material. There is increased expression of AO enzymes in response to OS. Formation of ROS results in activation of cell signaling pathways and induction of apoptosis. Silica can also give rise to RNS, mitochondrial dysfunction, and increased gene expression. Other papers also link silica intoxication to ROS and RNS (Porter et al. 2002; Castranova 2004) and cell signaling (Castranova 2004). A study of Chinese workers provides evidence that silica alone does not cause lung cancer, but must be enjoined with other contaminants that also play an important role (Chen et al. 2007).

3.4.5 Persulfate

Occupational asthma has been diagnosed from exposure to persulfate salts (Munoz et al. 2004). It is plausible to attribute toxicity to the oxidizing property of the toxin.

3.4.6 Perchlorate

Yang et al. (2005) reported induction of pulmonary fibrosis and acute inflammatory reaction by this oxidant. The level of malondialdehyde, a common indicator of OS from ROS, was elevated.

3.4.7 Paraquat

This effective herbicide usually enters the body by ingestion (Witschi 1997). Pulmonary toxicity evidently results from uptake by and accumulation in the lung. The mechanism of paraquat action in the lung is related to redox cycling involving oxygen, followed by formation of superoxide, a precursor for other ROS. There is evidence that lipid peroxidation is a contributing event in the pathogenesis of paraquat-induced lung lesions. Structurally, substance **18** belongs to the iminium (pyridinium) ET family. Cytotoxicity from paraquat was alleviated by AOs (Ikeda et al. 2003) and potentiated by hyperoxia (Leikauf and Driscoll 1993). Cellular conversion to paraquat radical was observed. In addition, there are other reports that suggest a link to generation of OS (Adachi et al. 2003; Cho et al. 2005).

3.4.8 Insecticides

Barthel (1981) found increased risk of lung cancer among agricultural workers exposed to insecticides and other pesticides. A positive correlation between duration of employment and mortality from lung cancer suggested a dose–effect relationship. Mechanisms of insecticide toxicity have been reviewed with focus on ET–ROS–OS (Kovacic 2003).

3.4.9 Tobacco

Lung cancer from cigarette smoking is a major cause of mortality. The diversity of carcinogens in tobacco, many of which are volatilized during the smoking process, gives rise to difficulty in pinpointing what mechanisms are operable and which are the causative agents. Among known toxicants in tobacco are N-nitrosoamines, PAHs, primary aromatic amines, nicotine, hydrogen peroxide, superoxide, and semiquinones (Kovacic and Jacintho 2001a). There is appreciable evidence for involvement of ROS in tobacco-induced toxicity, partly based on tobacco toxins known to decrease AOs, as well as the protection against toxicity afforded by AOs. Kim et al. (2004) suggest that chronic environmental tobacco smoke exposure can increase lipid peroxidation in the lung and red blood cells. The biochemistry of tobacco constituents and link to ET–ROS–OS are discussed elsewhere (Kovacic and Jacintho 2001a).

3.4.10 Cocaine

The abused drug, cocaine, has attracted considerable attention because of the toxicity and pulmonary complications it induces after inhalation (Nistal de Paz et al. 1984); such effects include severe changes in capillaries, alveoli, and bronchioles (Barroso-Moguel et al. 1999), as well as increased lung permeability

(Susskind et al. 1991). Moreover, cocaine-induced pulmonary hypertension can result in pulmonary edema (Lang and Maron 1991), and massive overdoses increased lung water retention and caused ascites (Robin et al. 1989). Antenatal administration is associated with fetal hypoxemia with apparent increased free radical production, as indicated by an observed decrease in GSH peroxidase activity (Sosenko 1993). Kovacic (2005) addressed the role of oxidative metabolites in toxicity and addiction in relation to ET and ROS. One of the important cocaine metabolites is a norcocaine nitroxide radical whose functionality is depicted as **19**.

3.4.11 Naphthalene

This aromatic hydrocarbon, a ubiquitous environmental contaminant, produces cytotoxicity in bronchiolar epithelial cells (Baldwin et al. 2005). One metabolic route entails formation of a 1,2-epoxide that serves as precursor to other products. The reactive intermediates deplete GSH, covalently bind to proteins, and cause lung necrosis (Phimister et al. 2005). GSH loss is believed to be a major determinant of inhaled reparatory toxicity (Phimister et al. 2004). Other reports show metabolism to 1,2- and 1,4-diols, as well as the corresponding quinones, which can undergo redox cycling as with **1** and **2**, to produce ROS (Kovacic and Somanathan 2007).

3.4.12 Polycyclic Aromatic Hydrocarbons

The PAH toxins are present in combustion-derived particulate matter and in cigarette smoke (Reed et al. 2003). Most attention has centered on the diolepoxide metabolite, which is an alkylator. However, based on a study of benzo[a]pyrene, other oxidative products are formed from PAHs, mainly quinones that can generate ROS by redox cycling. More information on this topic is available in a recent book chapter (Kovacic and Somanathan 2007).

3.4.13 Phthalates

Hoppin et al. (2004) report that exposure to the phthalate plasticizers is widespread, producing adverse respiratory outcomes in children. A report on phthalate metabolism reveals aromatic hydroxylation to catechol-type products, such as **20**; these metabolites have potential for redox cycling with accompanying production of ROS (Kovacic and Jacintho 2001b).

3.4.14 Quinones

Protein targets for 1,4-benzoquinone (**2**) and 1,4-naphthaquinone were the focus of attention in studies with human bronchial epithelial cells (Lame et al. 2003). Many aspects of their toxicity have been attributed to covalent modification of proteins.

The reviews on quinone organ toxicity addressed in this paper's introduction documents many examples of insults inflicted by ROS–OS that arise from redox transformations entailing ET.

3.4.15 Nitroaromatic Compounds

Results with 1-nitropyrene, an environmental pollutant, indicate that lung tissue is capable of producing both oxidative and reductive mutagenic metabolites, related to the structures **3** and **4**, several of which were more potent than the parent compound (King et al. 1984). Bond (1983) found that 6-nitrochrysene, a carcinogen, undergoes metabolic activation by enzymes in the hamster lung. Bond (1983) addresses the metabolism of aromatic nitro compounds in relation to ET-ROS-OS.

3.5 Other (Radiation)

Radiation therapy, an important modality in treatment of lung tumors, often causes accompanying injury believed to be a consequence of OS and the cascade of cytokine activity (Vujaskovic et al. 2002). A novel SOD mimetic demonstrates a significant protective effect. Details of the toxic mechanism are available elsewhere (Kovacic and Osuna 2000).

4 Illness and Oxidative Stress

The prior sections provide abundant evidence for involvement of ET-ROS-OS in toxic lung effects. It is not surprising, therefore, that a similar connection has been noted for various lung illnesses (Halliwell and Gutteridge, 1999). Several of the prominent pulmonary illnesses are briefly presented later.

4.1 Asthma

Andreadis et al. (2003) reported that asthma affects over 15 million people in the USA and causes 5,500 deaths yearly. Airway inflammation is associated with increased production of ROS and RNS, with resultant toxic reactions. Loss of protective AO defenses exacerbates the condition. Li et al. (2003) reviewed particulate air pollutants and the role of OS in asthma.

4.2 Chronic Obstructive Pulmonary Disease

COPD is a major and increasing global health problem and is predicted to become the third most common cause of death by 2020 (Barnes et al. 2003). This 2003 review addresses molecular and cellular mechanisms, including OS, inflammation emphysema, and bronchiolitis. Cigarette smoking is the most common cause of COPD. Rahman et al. (2000) confirm decreased AO capacity in patients with COPD and in smokers, indicating the role of OS.

5 Antioxidant Benefits and Pulmonary Toxicity

The extensive literature that exists on AOs illustrates the protective effect against lung insult of various AO types. Such protection supports the concept that ROS play an important role in lung toxicity. From the practical standpoint AOs can potentially prevent and ameliorate toxic injury. Romieu and Trenga (2001) reveal that the major AO-related defense systems are radical-scavenging agents that are present in intra and extracellular regions of the lung. The AOs include endogenous enzyme systems and nonenzymatic compounds. Among the various AOs investigated are thiols (Romieu and Trenga 2001; Sadowska et al. 2007; Rahman and MacNee 2000; Hagiwara et al. 2000; Pluzhnikov et al. 2004), vitamin C (Dietrich et al. 2002), vitamin E (Daga et al. 2003), flavonoids (Tabak et al. 2001; Shaheen et al. 2001), and selenium (Shaheen et al. 2001). There is an extensive literature on AO mixtures including those available in the diet. Representative examples of such AO mixtures are provided by Rahman (2002), Comhair and Erzurum (2002), Smit (2001), Wedgwood and Black (2003), and MacNee (2005).

6 Summary

The atmosphere is replete with a mixture of toxic substances, both natural and manmade. Inhalation of toxic substances produces a variety of insults to the pulmonary system. Lung poisons include industrial materials, particulates from mining and combustion, agricultural chemicals, cigarette smoke, ozone, and nitrogen oxides, among a large number of other chemicals and environmental contaminants. Many proposals have been advanced to explain the mode of action of pulmonary toxicants. In this review we focus on mechanisms of pulmonary toxicity that involve ET, ROS, and OS. The vast majority of toxicants or their metabolites possess chemical ET functionalities that can undergo redox cycling. Such recycling may generate ROS that can injure various cellular constituents in the lung and in other tissues. ET agents include quinones, metal complexes, aromatic nitro compounds, and conjugated iminium ions. Often, these agents are formed metabolically from parent toxicants. Such metabolic reactions are often catalytic and require only small amounts of the offending material. Oxidative attack is commonly associated with lipid peroxidation and oxidation of DNA, and it may result in strand cleavage and 8-OH-DG production. Toxicity is often accompanied by depletion of natural AOs, which further exacerbates the toxic effect. It is not surprising that the use of AOs, both natural in fruits and vegetables, as well as synthetic, may provide protection from the adverse effects of toxicant exposure. The mechanistic framework described earlier is also applicable to some of the more prominent pulmonary illnesses, such as, asthma, COPD, and cancer.

Acknowledgment We are grateful to Angelica Ruiz and Thelma Chavez for editorial assistance.

References

Adachi J, Ishii K, Tomita M, Fujita T, Nurhantari Y, Nagasaki Y, Ueno Y (2003) Consecutive administration of paraquat to rats induces enhanced cholesterol peroxidation and lung injury. Arch Toxicol 77:353–357.
Ahamad, S, Ahamad A, White AW (2006) Puringeric signaling and kinase activation for survival in pulmonary oxidative stress and disease. Free Rad Biol Med 41:29–40.
Andreadis AA, Hazen SL, Comhair SAA, Erzurum SC (2003) Oxidative and nitrosative events in asthma. Free Rad Biol Med 35:213–225.
Attene-Ramos MS, Wagner ED, Gaskin HR, Plewa MJ (2007) Hydrogen sulfide induces direct radical-associated DNA damage. Mol Cancer Res 5:455–459.
Baldwin RM, Shultz MA, Buckpitt AR (2005) Bioactivation of the pulmonary toxicants naphthalene and 1-nitronaphthalene by rat CYP2F4. J Pharmacol Exp Ther 312:857–865.
Barchowsky A, O'Hara KA (2003) Metal-induced cell signaling and gene activation in lung disease. Free Rad Biol Med 34:1130–1135.
Barnes PJ, Shapiro SD, Pauwels RA (2003) Chronic obstructive pulmonary disease: Molecular and cellular mechanisms. Eur Respir J 22:672–688.
Barreiro E, de la Puente B, Minguella J, Corominas JM, Serrano S, Hussain SNA, Gea J (2005) Oxidative stress and respiratory muscle dysfunction in severe chronic obstructive pulmonary disease. Am J Respir Crit Care Med 171:1116–1124.
Barroso-Moguel R, Villeda-Hernández J, Méndez-Armenta M, Santamaría A, Galván-Arzate S (1999) Alveolar lesions induced by sytemic administration of cocaine to rats. Toxicol Lett 110:113–118.
Barthel E (1981) Increased risk of lung cancer in pesticide-exposed male agricultural workers. J Toxicol Environ Health 8:1027–1040.
Baskar R, Li L, Moore PK (2007) Hydrogen sulfide-induces DNA damage and changes in apoptotic gene expression in human lung fibroblast cells. FASEB J 21:247–255.
Battacharya R, Kumar P, Sachan AS (1994) Cyanide induced changes in dynamic pulmonary mechanics in rats. Indian J Physiol Pharmacol 38:281–284.
Bebok Z, Varga K, Hicks JK, Venglarik CJ, Kovacs T, Chen L, Hardiman KM, Collawn JF, Sorscher EJ (2002) Reactive oxygen-nitrogen species decrease cystic fibrosis transmembrane conductance regulator expression and cAMP-mediated Cl$^-$ secretion in airway epithelia. J Biol Chem 227:43041–43049.

Beck-Speier I, Dayal N, Karg E, Kondrad L, Schumann G, Schulz H, Semmler M, Takenaka S, Stettmaier K, Bors W, Ghio A, Samet JM, Heyder J (2005) Oxidative stress and lipid mediators induced in alveolar macrophages by ultrafine particles. Free Rad Biol Med 38:1080–1092.

Beeh KM, Beier J, Haas IC, Kornmann O, Micke P, Buhl R (2002) Glutathione deficiency of the lower respiratory tract in patients with idiopathic pulmonary fibrosis. Eur Resp J 19:1119–1123.

Beltron R, Lee A (2001) Respiratory disorders. In: Lee A (ed) Adverse drug reactions. Pharmaceutical Press, London, pp. 137–155.

Bond JA (1983) Bioactivation and biotransformation of 1-nitropyrene in liver, lung and nasal tissue of rats. Mutat Res 124:315–324.

Bowler RP, Crapo JD (2002) Oxidative stress in allergic respiratory diseases. J Allergy Clin Immunol 110:345–356.

Boysen G, Georgieva NI, Upton PB, Jayaraj K, Li Y, Walker VE, Swenberg JA (2004) Analysis of diepoxide-specific cyclic N-terminal globin adducts in mice and rats after inhalation exposure to 1,3-butadiene. Cancer Res 64:8517–8520.

Briede JJ, Godschalk RWL, Emans MTG, de Kok TMC, van Agen E, van Maanen JMS, van Schooten F-J, Kleinjans JCS (2004) In vitro and in vivo studies on free radical and DNA adduct formation in rat lung and liver during benzo[a]pyrene metabolism. Free Rad Res 38:995–1002.

Brown DM, Donaldson K, Borm PJ, Schins PR, Denhart M, Gilmour P, Jimenez LA, Stone V (2004a) Calcium and reactive oxygen species-mediated activation of transcription factors and TNFa cytokine gene expression in macrophages exposed to ultrafine particles. Am J Physiol 286:L344–L353.

Brown LAS, Harris FL, Ping X-D, Gauthier TW (2004b) Chronic ethanol ingestion and the risk of acute lung injury: A role for glutathione availability? Alcohol 33:191–197.

Casalino-Matsuda SM, Monzon ME, Conner GE, Salathe M, Forteza RM (2004) Role of hyaluronan and reactive oxygen species in tissue kallikrein-mediated EGF receptor activation in human airways. J Biol Chem 279:21606–21616.

Castranova V (2004) Signaling pathways controlling the production of inflammatory mediators in response to crystalline silica exposure: Role of reactive oxygen/nitrogen species. Free Rad Biol Med 37:916–925.

Chang LW, Lo W-S, Lin P (2005) Trans, trans-2,4-decadienal, a product found in cooking oil fumes, induces cell proliferation and cytokine production due to reactive oxygen species in human bronchial epithelial cells. Toxicol Sci 87:337–343.

Chen F, Vallyathan V (2005) Molecular mechanisms of asbestos- and silica-induced lung cancer. In: Bachi D, Preuss HG (eds) Phytopharmaceuticals in cancer chemoprevention. CRC Press, Boca Raton, FL, pp. 41–62.

Chen WJ, Chi EY, Smuckler EA (1977) Carbon tetrachloride-induced changes in mixed function oxidases and microsomal cytochromes in the rat lung. Lab Invest 36:388–394.

Chen SJ, Wang JL, Chen JH, Huang RN (2002) Possible involvement of glutathione and p53 in trichloroethylene-and perchloroethylene-induced lipid peroxidation and apoptosis in human lung cancer cells. Free Rad Biol Med 15:464–472.

Chen W, Bochmann F, Sun Y (2007) Effects of work related confounders on the association between silica exposure and lung cancer: A nested case-control study among Chinese miners and pottery workers. Int Arch Occup Environ Health 80:320–326.

Cho H-Y, Kleeberger SR (2007) Genetic mechanisms of susceptibility to oxidative lung injury in mice. Free Rad Biol Med 42:433–445.

Cho JH, Yang DK, Kim L, Ryu JS, Lee HL, Lim CM, Koh YS (2005) Inhaled nitric oxide improves the survival of the paraquat-injured rats. Vasc Pharmacol 42:171–178.

Comhair SAA, Erzurum SC (2002) Antioxidant responses to oxidant-mediated lung diseases. Am J Physiol Lung Cell Mol Physiol 283:L246–L255.

Cruzan G, Carlson GP, Turner M, Mellert W (2005). Ring-oxidized metabolites of styrene contribute to styrene-induced Clara-cell toxicity in mice. J Environ Toxicol Environ Health A 68:229–237.

Csanady GA, Kessler W, Hoffmann HD, Filser JG (2003) A toxicokinetic model for styrene and its metabolite styrene-7,8-oxide in mouse, rat and human with special emphasis on the lung. Toxicol Lett 138:75–102.

Daga MK, Chhabra R, Sharma B, Mishra TK (2003) Effects of exogeneous vitamin E supplementation on the levels of oxidants and antioxidants in chronic obstructive pulmonary disease. J Biosci 28:7–11.

Dazy A-C, Auger F, Bailbe D, Blouquit S, Lombet A, Marano F (2003) The toxicity of H_2O_2 on the ionic homeostasis of airway epithelial cells in vitro. Toxicol In Vitro 17:575–580.

Dietrich M, Block G, Hudes M, Morrow JD, Norkus EP, Traber MG, Cross CE, Packer L (2002) Antioxidant supplementation decreases lipid peroxidation biomarker F_2-isoprostanes in plasma of smokers. Cancer Epidemiol Biomarkers Prev 11:7–13.

Dobson AW, Weber S, Dorman DC, Lash LK, Erikson KM, Aschner M (2003) Oxidative stress is induced in the rat brain following repeated inhalation exposure to manganese sulfate. Biol Trace Elem Res 93:113–125.

Donaldson K, Borm P (2007) Particle toxicology. CRC Press/Taylor & Francis, Boca Raton, FL, pp. 1–434.

Donaldson K, Jimenez LA, Rahman I, Faux SP, MacNee W, Gilmour PS, Borm PJA, Schins RPF, Shi T, Stone, V (2004) Resiratory health effects of ambient air pollution particles: Role of reactive species. Lung Biol Health Dis 187:257–288.

Dorman DC, Moulin FJ, McManus BE, Mahle KC, James RA, Struve MF (2002) Cytochrome oxidase inhibition induced by acute hydrogen sulfide inhalation: Correlation with tissue sulfide concentration in the rat brain, liver, lung, and nasal epithelium. Toxicol Sci 65:18–25.

Durak, I, Güven T, Birey M, Oztürk HS, Kurtipek O, Yel M, Dikmen B, Canbolat O, Kavutcu M, Kaçmaz M (1996) Halothane hepatotoxicity and hepatic free radical metabolism in guinea pigs: The effects of vitamin E. Can J Anaesth 43:741–748.

Elsayed NM, Omaye ST (2004) Biochemical changes in mouse lung after subcutaneous injection of the sulfur mustard 2-chlroethyl-4- chlorobutyl sulfide. Toxicol 199:195–206.

Emmler J, Hermanns MI, Steinritz D, Kreppel H, Kirkpatrick CJ, Bloch W, Szinicz L (2007) Assessment of alterations in barrier functionality and induction of proinflammatory and cytotoxic effects after sulfur mustard exposure of an in vitro coculture model of the human alveolo-capillary barrier. Inhal Toxicol 19:657–665.

Ergonul Z, Erdem A, Balkanci ZD, Kilinc K (2007) Vitamin E protects against lipid peroxidation due to cold-SO_2 coexposure in mouse lung. Inhal Toxicol 19:161–168.

Erikson KM, Dorman DC, Lash LH, Dobson AW, Aschner M (2004) Airborne manganese exposure differently affects end point of oxidative stress in age- and sex- dependent manner. Biol Trace Elem Res 100:49–62.

Fakhrzadeh L, Laskin JD, Gardner CR, Laskin DL (2004a) Superoxide dismutase-overexpressing mice are resistant to ozone-induced tissue injury and increases in nitric oxide and tumor necrosis factor-α. Respir Cell Mol Biol 30:280–287.

Fakhrzadeh L, Laskin JD, Laskin DL (2004b) Ozone-induced production of nitric oxide and TNF-α and tissue injury are dependent on NF-kB p50. Am Physiol Soc 287:L279–L285.

Fedan JS, Fedan KB, Hubbs AF (2006) Popcorn workers lung: In vitro exposure to diacetyl, an ingredient in microwave butter flavoring, increases reactivity to methacholine. Toxicol Appl Pharmacol 215:17–22.

Fubini B, Hubbard A (2003) Reactive oxygen species (ROS) and reactive nitrogen species (RNS) generation by silica in inflammation and fibrosis. Free Rad Biol Med 34:1507–1516.

Giovanetti A, Rossi L, Mancuso M, Lombardi CC, Marasco MR, Manna F, Altavista P, Massa EM (1998) Analysis of lung damage induced by trichoroethylene inhalation in mice fed diets with low, normal, and high copper content. Toxicol Pathol 26:628–635.

Gonzales-Flecha B (2004) Oxidant mechanisms in response to ambient air particles. Mol Aspects Med 25:169–182.

Grasemann H, Tschiedel E, Groch M, Klepper J, Ratjen F (2007) Exhaled nitic oxide in chlorine after accidental exposure to chlorine gas. Inhal Toxicol 19:895–898.

Green FH, Schürch S, de Sanctis, GT, Wallace JA, Cheng S, Prior M (1991) Effects of hydrogen sulfide exposure on surface properties of lung surfactant. J Appl Physiol 70:1943–1949.

Hadjiivanova NB, Salovski PZ, Groseva MM, Charakchieva SB, Nechev CK (1987) Early effects of n-hexane and irradiation on the lung surfactant system. Acta Physiol Pharmacol Bulg 13:25–29.

Hagiwara S-I, Ishii Y, Kitamura S (2000) Aerosolized administration of N-acetylcysteine attenuates lung fibrosis induced by bleomycin in mice. Am J Respir Crit Care Med 162:225–231.

Halliwell B, Gutteridge JMC (1999) Free radicals in biology and medicine. Oxford University Press, New York, (a) pp. 1–897, (b) pp. 679–684, (c) pp. 27, 581, (d) p. 576, (e) pp. 27, 55, 86, 91.

Hammerschmidt S, Wahn H (2004) The oxidants hypochlorite and hydrogen peroxide induce distinct patterns of acute lung injury. Biochim Biophys Acta Mol Basis Dis 1690:258–264.

Henderson RF, Belinsky SA (1993) Biological markers of respiratory tract exposure. In: Gardner DE, Crapo JD, McClellan RO (eds) Toxicology of the lung. Raven Press, New York, pp. 253–282.

Hessel PA, Herbert FA, Melenka LS, Yoshida K, Nakaza M (1997) Lung health in relation to hydrogen sulfide exposure in oil and gas workers in Alberta, Canada. Am J Ind Med 31:554–557.

Hickman-Davis JM, Fang FC, Nathan C, Shepherd VL, Voelker DR, Wright JR (2001) Lung surfactant and reactive oxygen-nitrogen species: Antimicrobial activity and host–pathogen interactions. Am J Physiol Cell Mol Physiol 281:L517–L523.

Hong HH, Houle CD, Ton TV, Sills RC (2007) K-ras mutations in lung tumors and tumors from other organs are consistent with a common mechanism of ethylene oxide tumorigenesis in the B6C3F1 mouse. Toxicol Pathol 35:81–85.

Hoppin JA, Ulmer R, London SJ (2004) Phthalates exposure and pulmonary function. Environ Health Perspect 112:571–574.

Ilgazli A, Sengul C, Maral H, Ozden M, Ercin C (2004) The effects of thinner inhalation on superoxide dismutase activities, malondialdehyde and glutathione levels in rat lungs. Clin Chim Acta 343:141–144.

Ikeda K, Kumagai Y, Nagano Y, Matsuzawa N, Kojo S (2003) Change in the concentration of vitamins C and E in rat tissues by paraquat administration. Biosci Biotechnol Biochem 67:1130–1131.

Ito K, Sumiko H, Yusuke K, Kawaishi S (2005) Mechanism of site-specific DNA damage induced by ozone. Mutat Res Gen Toxicol Environ Mutagen 585:60–70.

Jacintho JD, Kovacic P (2003) Neurotransmission and neurotoxicity by nitric oxide, catecholamines and glutamate: Unifying themes of reactive oxygen species and electron transfer. Curr Med Chem 10: 2693–2704.

Jurczyk AS, Barzdo M, Jankowska B, Meissner E, Berent J, Kordel K, Szram S (2003) Influence of selected alcohols on oxidative stress parameters in rat lungs. Z Zagadnien Nauk Sadowych 55:50–59.

Kaufmann W, Mellert W, van Ravenzwaay B, Landsiedel R, Poole A (2005) Effects of styrene and its metabolites on different lung compartments of the mouse-cell proliferation and histomorphology. Regul Toxicol Pharmacol 42:24–36.

Kelly FJ, Mudway IS (2003) Protein oxidation at the air–lung interface. Amino Acids 25:375–396.

Kim D-H, Suh Y-S, Mun K-C (2004) Tissue levels of malondialdehyde alter passive smoke exposure of rats for a 24-week period. Nicotine Tob Res 6:1039–1042.

King LC, Kohan MJ, Ball LM, Lewtas J (1984) Mutagenicity of 1-nitropyrene metabolites from lung-S9. Cancer Lett 22:255–262.

Knaapen AM, Borm PJA, Albrecht C, Schins RPF (2004) Inhaled particles and lung cancer, Part A: Mechanisms. Int J Cancer 109:799–809.

Kovacic P (2003) Mechanism of organophosphates (nerve gases and pesticides) and antidotes: Electron transfer and oxidative stress. Curr Med Chem 10:2705–2710.

Kovacic P (2005) Role of oxidative metabolites of cocaine in toxicity and addiction: Oxidative stress and electron transfer. Med Hypotheses 64:350–356.

Kovacic P (2006) Novel electrochemal approach to enhanced toxicity of 4-oxo-2-nonenal ve. 4-hydroxy-2-nonenal (role of imine): Oxidative stress and therapeutic modalities. Med Hypotheses 67:151–156.

Kovacic P (2007) What is the basic cause of diacetyl toxicity in popcorn lung disease? Possible remedies, submitted.

Kovacic P, Becvar LE (2000) Mode of action of anti-infective agents: Emphasis on oxidative stress and electron transfer. Curr Pharm Des 6:143–167.

Kovacic P, Cooksy AL (2005) Unifying mechanism for toxicity and addiction by abused drugs: Electron transfer and reactive oxygen species. Med Hypotheses 64:366–367.

Kovacic P, Jacintho JD (2001a) Mechanisms of carcinogenesis: Focus on oxidative stress and electron transfer. Curr Med Chem 8:773–796.

Kovacic P, Jacintho JD (2001b) Reproductive toxins: Pervasive theme of oxidative stress and electron transfer. Curr Med Chem 8:863–892.

Kovacic P, Jacintho JD (2003) Systemic lupus erythematosus and other autoimmune diseases from endogenous and exogenous agents: Unifying theme of oxidative stress. Mini Rev Med Chem 3:568–575.

Kovacic P, Osuna JA (2000) Mechanisms of anticancer agents: Emphasis on oxidative stress and electron transfer. Curr Pharm Des 6:277–309.

Kovacic P, Pozos RS (2006) Cell signaling (mechanism and reproductive toxicity): Redox chains, electrons, relays, conduit, electrochemistry, and other medical implications. Birth Defects Res C 8:333–344.

Kovacic P, Somanathan R (2005) Neurotoxicity: The broad framework of electron transfer, oxidative stress and protection by antioxidants. Curr Med Chem Cent Nerv Syst Agents 5:249–258 (and references therein).

Kovacic P, Somanathan R (2006a) Beneficial effects of antioxidants in relation to carcinogens, toxins and various illnesses. In: Panglossi HV (ed) Frontiers in antioxidants research. Nova Science, Hauppauge, NY, Ch. 1, pp. 1–38.

Kovacic P, Somanathan R (2006b) Alcohol mechanisms, cell signaling, toxicity, addiction, prevention, therapy and beneficial effects. In: Brozner EY (ed) New research on alchohol abuse and alcoholism. Nova Science, Hauppauge, NY, Ch. 4, pp. 40–101.

Kovacic P, Somanathan R (2007) Mechanism of tumorigenesis: Focus on oxidative stress, electron transfer and antioxidants. In: Wong DK (ed), Tumorigenesis research advances. Nova Science, Hauppauge, NY, Ch. 2.

Kovacic P, Thurn LA (2005) Cardiovascular toxins from the perspective of oxidative stress and electron transfer. Curr Vasc Pharmacol 3:107–117.

Kovacic P, Sacman A, Wu-Weis M (2002) Nephrotoxins: Widespread role of oxidative stress and electron transfer. Curr Med Chem 9:823–847.

Kovacic P, Pozos, RS, Somanathan R, Shangari R, O'Brien PJ (2005) Mechanism of mitochondrial uncouplers, inhibitors, and toxins: Focus on electron transfer, free radicals, and structure–activity relationships. Curr Med Chem 5:2601–2623.

Lai C-H, Liou S-H, Lin H-C, Shih T-S, Tsai P-J, Chen J-S, Yang T, Jaakkola JJK, Stickland PT (2005) Exposure to traffic exhausts and oxidative DNA damage. Occup Environ Med 62:216–222.

Lame MW, Jones AD, Wilson DW, Segall HJ (2003) Protein targets of 1,4-benzoquinone and 1,4-naphthaquinone in human bronchial epithelial cells. Proteomics 3:479–495.

Lang SA, Maron MB (1991) Hemodynamic basis for cocaine-induced pulmonary edema in dogs. J Appl Physiol 71:1166–1170.

Langen RCJ, Korn SH, Wouters EFM (2003) ROS in the local and systemic pathogenesis of COPD. Free Rad Biol Med 35:226–235.

Laskin DL, Gardner CR, Gerecke DR, Laskin JD (2004) Ozone-induced lung injury: Role of macrophages and inflammatory mediators. Lung Biol Health Dis 187:289–316.

Lee KS, Kim SR, Park SJ, Park HS, Min KH, Jin SM, Lee MK, Kim UH, Lee YC (2006) Peroxisome proliferators activated receptor-γ modulates reactive oxygen species generation and activation of nuclear factor-kB and hypoxia-inducible factor 1α in allergic airway disease of mice. J Allergy Clin Immunol 118:120–127.

Leikauf G, Driscoll K (1993) Cellular approaches in respiratory tract toxicology. In: Gardner DE, Crapo JD, McClellan RO (eds) Toxicology of the lung. Raven Press, New York, pp. 335–370.

Levitt JD (1975) The biochemical basis of anesthetic toxicity. Surg Clin North Am 55:801–818.
Lewis AB, Taylor MD, Roberts JR, Leonard SS, Shi X, Antonini JM (2003) Role of metal-induced reactive oxygen species generation in lung responses caused by residual oil fly ash. J Biosci 28:13–18.
Li N, Hao M, Phalen RF, Hinds WC, Nel AE (2003) Particulate air pollutants and asthma. A paradigm for the role of oxidative stress in PM-induced adverse health effects. Clin Immunol 109:250–265.
Liang X, Wang P-H (2004) Phosgene-induced oxidative injury in rats and effects of $NaHCO_3$ buffer. Disi Jumji Daxue Xuebao 25:1235–1237.
Lungarella G, Barni-Comparini I, Fonzi L (1984) Pulmonary changes in rabbits by long-term exposure to n-hexane. Arch Toxicol 55:224–228.
Luo JC, Cheng TJ, Kuo HW, Chang MJ (2004) Decreased lung function associated with occupatoional exposure to epichlorohydrin and the modification effects of glutathione *S*-transferase polymorphism. J Occup Environ Med 46:280–286.
Machado RF, Nerker, M-VL, Dweik RA, Hammel J, Janocha A, Pyle J, Laskowski D, Jennings C, Arroliga AA, Erzurum SC (2004) Nitric oxide and pulmonary arterial pressures in pulmonary hypertension. Free Rad Biol Med 37:1010–1017.
MacNee W (2005) Pulmonary and systemic oxidant/antioxidant imbalance in chronic obstructive pulmonary disease. Proc Am Thorac Soc 2:50–60.
Manoury B, Nenan S, Leclerc O, Guenon I, Boichot E, Planquois J-M, Bertrand CP, Lagente V (2005) The absence of reactive oxygen species production protects mice against bleomycin-induced pulmonary fibrosis. Respir Res 6:11.
Mazzullo M, Colacci A, Grilli S, Prodi G, Arfellini G (1984) *In vivo* and *in vitro* binding of epichlorohydrin to nucleic acids. Cancer Lett 23:81–90.
Meier BW, Gomez JD, Kirichenko OV, Thompson JA (2007) Mechanistic basis for inflammation and tumor promotion in lungs of 2,6-*di-tert*-butyl-4-methylphenol-treated mice: Electrophilic metabolites alkylate and inactivate antioxidant enzymes. Chem Res Toxicol 20:199–207.
Meng Z, Liu Y (2007) Cell morphological ultrastructural changes in various organs from mice exposed by inhalation to sulfur dioxide. Inhal Toxicol 19:543–551.
Meng Z, Qin G, Zhang, B, Geng H, Bai Q, Bai W, Liu, C (2003) Oxidative damage of sulfur dioxide inhalation on lungs and hearts of mice. Environ Res 93:285–292.
Mikhaĭlova A, Petrova S, Donchev N (1987) Biochemical and histological research on lung tissue in experimental carbon disulfide exposure. Probl Khig 12:103–108.
Miller FJ, Overton JH, Kimbell JS, Russell ML (1993) Regional respiratory tract absorption of inhaled reactive gases. In: Gardner DE, Crapo JD, McClellan RO (eds) Toxicology of the lung. Raven Press, New York, pp. 485–525.
Mossman BT, Gee JBL (1993) Pulmonary reactions and mechanisms of toxicity of inhaled fibers. In: Gardner DE, Crapo JD, McClellan RO (eds) Toxicology of the lung. Garden Press, New York, pp. 371–387.
Munoz X, Cruz MJ, Orriols R, Torres F, Espuga M, Morell F (2004) Validation of specific inhalation challenge for the diagnosis of occupational asthma due to persulfate salts. Occup Environ Med 61:861–866.
Neal RA, Halpert J (1982) Toxicology of thiono-sulfur compounds. Annu Rev Pharmacol Toxicol 22:321–339.
Nistal de Paz F, Fernández JJO, González JA, Colubi LC (1984) Pulmonary complications related to cocaine consumption. Ann Med Int 16:371–379.
Ortega R, Fayard B, Salomé M, Devés G, Susini J (2005) Chromium oxidation state imaging in mammalian cells exposed in vitro to soluble or particulate chromate compounds. Chem Res Toxicol 18:1512–1519.
Pääkkö P, Anttila S, Sormunen R, Ala-kokko L, Peura, R, Ferrans VJ, Ryhänen L (1996) Biochemical and morphological characterization of carbon tetrachloride-induced lung fibrosis in rats. Arch Toxicol 70:540–552
Pant SC, Vijayaraghavan R, Das Gupta S (1993) Sarin induced lung pathology and protection by standard therapy regime. Biomed Environ Sci 6:103–111.

Phimister A J, Lee MG, Morin D, Buckpitt AR, Plopper CG (2004) Glutathione depletion is a major determinant of inhaled naphthalene respiratory toxicity and naphthalene metabolism in mice. Toxicol Sci 82:268–278.

Phimister AJ, Nagasawa HT, Buckpitt AR, Plopper CG (2005) Prevention of naphthalene-induced pulmonary toxicity by glutathione prodrugs: Roles for glutathione depletion in adduct formation and cell injury. J Biochem Mol Toxicol 19:42–51.

Pinho RA, Chiesa D, Mezzomo KM, Andrades ME, Bonatto F, Gelain D, Pizzol F, Knorst MM, Moreira JC (2002) Oxidative stress in chronic obstructive pulmonary disease patients submitted to rehabilitation program. Resp Med 101:1830–1835.

Pluzhnikov N, Tyaptin A, Zemlyanoy A, Varlashova M, Torkounov P, Lupachyov Yu (2004) The state of antioxidant system in brain of rats in the toxic pulmonary edema. Biomed Khim 50:57–63

Poli G, Cheesman KH, Dianzani MU, Slater TF (1989) Free Radicals in the pathogenesis of liver injury. Pergamon, New York, pp. 1–330.

Porter DW, Millecchia L, Robinson VA, Hubbs A, Willard P, Pack D, Ramsay D, McLaurin J, Khan A, Landsittel D, Teass A, Castranova V (2002) Enhanced nitric oxide and reactive oxygen species production and damage after inhalation of silica. Am J Physiol Lung Cell Mol Physiol 283:L485–L493.

Proudfoot AT (2003) Pentachlorophenol poisoning. Toxicol Rev 22:3–11.

Prows DR, McDowell SA, Aronow BJ, Leikauf GD (2003) Genetic susceptibility to nickel-induced acute lung injury. J Chemosphere 51:1139–1148.

Pryor WA, Houk KN, Foote CS, Fukuto JM, Ignarro LJ, Squadrito GL, Davies KJ (2006) Free radical biology and medicine: It's a gas, man! Free Rad Biol Med 291:R491–R511.

Puhakka A (2005) Nitric oxide synthases and reactive oxygen species damage in pleural and lung tissues and neoplasia. University of Helsinki, Oulu.

Pyykkö K (1983) Time-course of effects of toluene on microsomal enzymes in rat liver, kidney and lung during and after inhalation exposure. Chem Biol Interact 44:299–310.

Qiu R, Jiang Y, Fan W-C (2004) Mechanism of smoke nitrogen oxides on inhalation lung mediated by peroxynitrite. Xiaofang Kexue Yu Jishu Bianjibu 23:421–424.

Rahman I (2002) Oxidative stress and gene transcription in asthma and chronic obstructive pulmonary disease: Antioxidant therapeutic targets. Curr Drug Targets Inflamm Allergy 1:291–315.

Rahman I (2003) Oxidative stress, chromatin remodeling and gene transcription in inflammation and chronic lung diseases. J Biochem Mol Biol 36:95–109.

Rahman I, MacNee W (2000) Regulation of redox glutathione levels and gene transcription in lung inflammation: Therapeutic approaches. Free Rad Biol Med 28:1405–1420.

Rahman I, Swarska E, Henry M, Stolk J, MacNee W (2000) Is there any relationship between plasma antioxidant capacity and lung function in smokers and in patients with chronic obstructive pulmonary disease? Thorax 55:189–193.

Reed M, Monske M, Lauer F, Meserole S, Born J, Burchiel S (2003) Benzo[a]pyrene diones are produced by photochemical and enzymatic oxidation and induce concentration-dependent decreases in the proliferative state of human pulmonary epithelial cells. J Toxicol Environ Health A 66:1189–1205.

Rhoden CR, Lawrence J, Godleski JJ, Gonzalez-Flecha B (2004) N-Acetylcysteine prevents lung inflammation after short-term inhalation exposure to concentrated ambient particles. Toxicol Sci 79:296–303.

Risch A, Ramroth H, Raedts V, Rajaee-Behbahani N, Schmezer P, Bartsch H, Becher, H, Dietz A (2003) Laryngeal cancer risk in Caucasians is associated with alcohol and tobacco consumption but not modified by genetic polymorphisms in class I alcohol dehydrogenases ADH1B and ADH1C, and glutathione-S-transferases GSTM1 and GSTT1. Pharmacogenetics 13:225–230.

Robin ED, Wong RJ, Ptashne KA (1989) Increased lung water and ascites after massive cocaine overdosage in mice and improved survival related to beta-adrenergic blockage. Ann Intern Med 110:202–207.

Romieu I, Trenga C (2001) Diet and obstructive lung diseases. Epidemiol Rev 23:268–287.

Ryter SW, Otterbein LE (2004) Carbon monoxide in biology and medicine. Bioessays 26:270–280.

Sadowska AM, Manuel-Keenoy B, De Backer WA (2007) Antioxidant and anti-inflammatory efficacy of NAC in the treatment of COPD: Discordant in vitro and in vivo dose-effects: A review. Pulmon Pharmcol Ther 20:9–22.

Sahu SC, Lowther DK, Jones SL (1982) Biochemical response of rat lung to inhaled *n*-hexane. Toxicol Lett 12:13–17.

Salnikow K, Su W, Blagosklonny MV, Costa M (2000) Carcinogenic metals induce hypoxia-inducible factor-stimulated transcription by reactive oxygen species independent mechanism. Canc Res 60:3375–3378.

Schmiederer M, Knutson E, Muganda P, Albrecht T (2005) Acute exposure of human lung cells to 1.3-butadiene diepoxide results in G1 and G2 cell cycle arrest. Environ Mol Mutagen 45:354–364.

Sciuto AM, Cascio MBM, Moran TS, Forster JS (2003) The fate of antioxidant enzymes in bronchoalveolar lavage fluid over 7 days in mice with acute lung injury. Inhal Toxicol 15:675–685.

Seguy N, Hildebrand HF, Haguenoer JM (1994) Toxic action of ethylene oxide on pulmonary cells (L132) cultured under aerobic conditions. Toxicol Lett 70:23–32.

Shaheen SO, Sterne JAC, Thompson RL, Songhurst CE, Margetts BM, Burney,PGJ (2001) Dietery antioxidants and asthma in adults. Am J Crit Care Med 164:1823–1828.

Shi H, Hudson LG, Ding W, Wang S, Cooper KL, Liu S, Chen Y, Shi X, Liu KJ (2004) Arsenite causes DNA damage in keratinocytes via generation of hydroxyl radicals. Chem Res Toxicol 17:871–878.

Shukla A, Gulumian M, Hei TK, Kamp D, Rahman Q, Mossman BT (2003a) Multiple roles of oxidants in the pathogenesis of asbestos-induced diseases. Free Rad Biol Med 34:1117–1119.

Shukla A, Ramos-Nino M., Mossman B (2003b) Cell signaling and transcription factor activation by asbestos in lung injury and disease. Int J Biochem Cell Biol 35:1198–1209.

Smit HA (2001) Chronic obstructive pulmonary disease, asthma and protective effects of food intake: From hypothesis to evidence? Respir Res 2:261–264.

Sosenko IR (1993) Antenatal cocaine exposure produces accelerated surfactant maturation without stimulation of antioxidant enzyme development in the late gestation rat. Pediatr Res 33:327–331.

Sugimoto R, Kumagai Y, Nakai Y, Ishii T (2005) 9,10-Phenanthroquinone in diesel exhaust particles down regulates Cu, Zn-SOD and HO-1 in human pulmonary epithelial cells: Intracellular iron scavenger 1,10-phenanthroline affords protection against apoptosis. Free Rad Biol Med 38:388–395.

Susskind H, Weber DA, Volkow ND, Hitzemann R (1991) Increased lung permeability following long-term use of free-base cocaine (crack). Chest 100:903–909.

Tabak C, Arts ICW, Smit HA, HeederikD, Kromhout D (2001) Chronic obstructive pulmonary disease and intake of catechins, flavonols, and flavones. Am J Respir Crit Care Med 164:61–64.

Tokunaga I, Gotohda T, Ishigami A, Kitamura, O, Kubo S-I (2003) Toluene inhalation induced 8-hydroxy-2′-deoxyguanosine formation as the peroxidative degeneration in rat organs. Legal Med 5:34–41.

Tomasi A, Billing S, Garner A, Slater TF, Albano E (1983) The metabolism of halothane by hepatocytes: A comparison between free radical spin trapping and lipid peroxidation in relation to cell damage. Chem Biol Interact 46:353–368.

Torkelson TR, Oyen F, Rowe VK (1976) The toxicity of chloroform as determined by single and repeated exposure of laboratory animals. Am Ind Hyg Assoc 37:697–705.

Turi JL, Yang F, Garrick MD, Piantadosi CA, Ghio, AJ (2004) The iron cycle and oxidative stress in the lung. Free Rad Biol Med 36:850–857.

Upadhyay D, Kamp DW (2003) Asbestos-induced pulmonary toxicity: Role of DNA damage and apoptosis. Exp Biol Med 228:650–659.

Valacchi G, Pagnin E, Corbacho AM, Olano E, Davis PA, Packer L, Cross CE (2004) In vivo ozone exposure induces antioxidant/stress-related responses in murine lung and skin. Free Rad Biol Med 36:673–681.

Valavanidis A, Fiotakis K, Bakea E, Vlahogianni T (2005) Electron paramagnetic resonance of the generation of reactive oxygen species catalyzed by transition metals and quinoid redox cycling by inhalable ambient particulate matter. Redox Rep 10:37–51.

van der Vliet A, Eiserich JP, Shigenaga MK, Cross CE (1999) Reactive nitrogen species and tyrosine nitration in the respiratory tract. Am J Respir Crit Care Med 160:1–9.

Van Helden HPM, Kuijpers WC, Diemel RV (2004) Asthmalike symptoms following intratracheal exposure of guinea pigs to sulfur mustard aerosol: Therapeutic efficacy of exogenous lung surfactant curosurf and salbutamol. Inhal Toxicol 16:537–548.

Vassilakopoulos T, Hussain, SNA (2007) Ventilatory muscle activation and inflammation: Cytokines, reactive oxygen species, and nitric oxide. J Appl Physiol 102:1687–1695.

Vettori MV, Caglieri A, Goldoni M, Catoldi AF, Dare E, Alinovi R, Ceccatelli S, Mutti A (2005) Analysis of oxidative stress in SK-N-MC neurons exposed to styrene-7,8-oxide. Toxicol In Vitro 19:11–20.

Vujaskovic Z, Batinic-Haberle I, Rabbani ZN, Feng Q-F, Kang SK, Spasojevic I, Samulski TV, Fridovich I, Dewhirst MW, Anscher MS (2002) A small molecular weight catalytic metalloporphyrin antioxidant with superoxide dismutase (SOD) mimetic properties protects lungs from radiation-induced injury. Free Rad Biol Med 33:857–863.

Waldman WJ, Kristovich R, Knight DA, Dutta PK (2007)Inflammatory properties of iron-containing carbon nanoparticles. Chem Res Toxicol 20:1149–1154.

Wang L, Medan D, Mercer R, Overmiller D, Leonard S, Castranova V, Shi X, Ding M, Huang C, Rojanasakul Y (2003) Vanadium-induced apoptosis and pulmonary inflammation in mice: Role of reactive oxygen species. J Cell Physiol 195:99–107.

Wang Q-S, Zheng Y-M, Dong L, Ho Y-S, Guo Z, Wang Y-X (2007) Role of mitochondrial reactive oxygen species in hypoxia-dependent increase in intracellular calcium in pulmonary artery myocytes. Free Rad Biol Med 42:642–653.

Watt BE, Proudfoot AT, Vale JA (2004) Hydrogen peroxide poisoning. Toxicol Rev 23:51–57.

Wedgwood S, Black SM (2003) Role of reactive oxygen species in vascular remodeling associated with pulmonary hypertension. Antioxid Redox Signal 5:759–769.

Willis RJ, Recknagal RO (1979) Potentiation by carbon tetrachloride of NADPH-dependent lipid peroxidation in lung microsomes. Toxicol Appl Pharmacol 47:89–94.

Witschi H (1997) Selected examples of free-radical mediated lung injury. In: Wallace KB (ed) Free radical toxicology. Taylor & Francis, London, pp. 279–293.

Woo C-H, Yoo M-H, You H-J, Cho S-H, Mun, Y-C, Seong C-M, Kim J-H (2003) Transepithelial migration of neutrophils in response to leukotriene B_4 is mediated by a reactive oxygen species-extracellular signal-regulated kinase-linked cascade. J Immunol 170:6273–6279.

Wu KY, Ranasinghe A, Upton PB, Walker VE, Swenbwerg JA (1999) Molecular dosimetry of endogenous and ethylene oxide-induced N7-(2-hydroxyethyl) guanine formation in tissues of rodents. Carcinogenesis 20:1787–1792.

Xi Z, Chao F, Yang, D, Sun Y, Li G, Zhan H, Zhan W, Yang Y, Liu H (2004) Experimental study of the DNA damage induced by formaldehyde. Huanjiing Kexue Xuebao 24:719–722.

Xia T, Korge P, Weiss JN, Li N, Venkatesen MI, Sioutas C, Nel A (2004) Quinones and aromatic chemical compounds in particulate matter induce mitochondrial dysfunction: Implications for ultrafine particle toxicity. Environ Health Perspect 112:1347–1358.

Xie J, Fan R (2007) Protein oxidation and DNA–protein crosslink induced by sulfur dioxide in lungs, livers, and hearts from mice. Inhal Toxicol 19:759–765.

Yang H-B, Peng K-L, Zhao S-L, Chu Q-L, Lu C-R, Liu Y, Liu C, Yang L (2005) Exploring the effects of ammonium perchlorate on pulmonary fibrosis. Huanjing Yu Zhiye Yixue 22:43–45.

Yang YG, Huang ZX, Cheng X (2006) Lung, liver and kidney impairment caused by inhalation of normal hexane. Chin J Ind Hyg Occup Dis 24:292–294.

Yin XJ, Ma JYC, Antonini JM, Castranova VM, Ma JKH (2004) Roles of reactive oxygen species and heme oxygenase-1 in modulation of alveolar macrophage-mediated pulmonary immune responses to *Listeria monocytogenes* by diesel exhaust particles. Toxicol Sci 82:143–153.

Yoon BI, Hirabayashi Y, Kawasaki Y, Tsuboi I, Ott T, Kodama Y, Kanno J, Kim DY, Willecke K, Inoue T (2004) Exacerbation of benzene pneumotoxicity in connexin 32 knockout mice: Enhanced proliferation of CYP2E1-immunoreactive alveolar epithelial cells. Toxicology 195:19–29.

Yu G, Liu J, Li T, Li L, Li W, Zhu J, Li X (2004) Effects of formaldehyde on lung histomorphology and level of lipid peroxide in mice. Jilin Daxue Xuebao Yixueban 30:888–889.

Yuan Z, Schellekens H, Warner L, Janssen-Heininger J, Burch P, Heintz NH (2003) Reactive nitrogen species block cell re-entry through sustained production of hydrogen peroxide. Am J Res Cell Mol Biol 26:705–712.

Zaher TE, Miller EJ, Morrow DMP, Javdan M, Mantell LL (2007) Hyperoxia-induced signal transduction pathways in pulmonary epithelial cells. Free Rad Biol Med 42:897–908.

Zhang Q, Sun S, Wang Z, Yang D (2004) Effects of formaldehyde on activities of antioxidase by subchronic inhalation in mice. Disi Junyi Daxue Xuebao 24:2206–2207.

Zhang S, Villalta PW, Wang M, Hecht S (2007) Detection and quantitation of acrolein-derived 1,N2-propanodeoxyguanosine adducts in human lung by liquid chromatography-electrospray ionization-tandem mass spectrometry. Chem Res Toxicol 20:565–571.

Risk Assessment of *Pseudomonas aeruginosa* in Water

Kristina D. Mena and Charles P. Gerba

Contents

1 Introduction 72
2 General Characteristics 72
3 Medical Importance 73
4 Health Effects in Humans 75
    4.1 Nosocomial Disease 76
    4.2 Septicemia 77
    4.3 Endocarditis 77
    4.4 Osteomyelitis 77
    4.5 Pneumonia 78
    4.6 Urinary Tract Infections 78
    4.7 Gastrointestinal Infections 79
    4.8 Burn Patients 80
    4.9 Skin Infections 80
    4.10 Ear Infections 81
    4.11 Eye Infections 82
    4.12 Meningitis 83
5 Special Populations 83
    5.1 Children 83
    5.2 AIDS Patients 85
    5.3 Granulocytopenic Patients 86
    5.4 Cancer Patients 86
6 Occurrence and Survival of *P. aeruginosa* 87
    6.1 Drinking Water 87
    6.2 Recreational Waters – Hot Tubs, Whirlpools, and Swimming Pools 88
    6.3 Surface Waters 89
    6.4 Biofilms 92
    6.5 *P. aeruginosa* from Other Sources 93
    6.6 Survival 94
7 Drinking Water Treatment for the Removal of *P. aeruginosa* 94
    7.1 Iodine, Free Chlorine, and Chloramine 94
    7.2 Other Disinfectants 96

K.D. Mena (✉)
University of Texas – Houston School of Public Health, Houston, Texas, USA
e-mail: kristina.d.mena@uth.tmc.edu

8	Epidemiological Evidence for Transmission by Water		97
	8.1	Hot Tubs, Spas, and Whirlpools	97
	8.2	Swimming Pools	99
	8.3	Tap Water	99
9	Water Quality Standards and Guidelines for *P. aeruginosa*		101
10	Risk Assessment		101
	10.1	Occurrence of *P. aeruginosa* in Drinking Water	102
	10.2	Infective Dose	102
	10.3	Appraisal of Risk	103
	10.4	Risks Associated with *P. aeruginosa* in Drinking Water	104
11	Data Gaps		105
12	Summary		105
References			107

1 Introduction

Pseudomonads are a large group of free-living bacteria that live primarily in soil, seawater, and fresh water. They also colonize plants and animals, and are frequently found in home and clinical settings. Pseudomonads are highly versatile and can adapt to a wide range of habitats, and can even grow in distilled water. This adaptability accounts for their constant presence in the environment. They have an extensive impact on ecology, agriculture, and commerce. They are responsible for food spoilage and degradation of petroleum products and materials. In agriculture, pseudomonads rank among the most important plant pathogens. In normal healthy humans, they are responsible for eye and skin diseases. They also cause serious life-threatening illnesses in burn and surgical patients and in immunocompromised individuals. Contamination of recreational waters and tap water has been associated with outbreaks of *Pseudomonas*; however, the relative role water plays in the transmission of this bacterium to humans is still unclear. The goal of this review is to assess existing literature on the potential risks associated with waterborne *Pseudomonas aeruginosa*.

2 General Characteristics

There are many species in the genus *Pseudomonas*; however, *P. aeruginosa* is the species most commonly associated with human illness. *P. aeruginosa* is most commonly associated with ear and skin infections acquired from contact with contaminated pools, hot tubs, and whirlpools. It causes serious nosocomial (hospital-acquired) infections in cystic fibrosis and burn patients, and immunocompromised individuals.

Based on comparative sequencing of the ribosomal RNA (ribonucleic acid), pseudomonads are classified in the *Proteobacteria* phylum. The *Proteobacteria* contain five clusters of bacteria, designated by the Greek letters Alpha, Beta, Gamma, Delta, and Epsilon. The genus *Pseudomonas* belongs to the Gamma 16s RNA ribosomal group (Kersters et al. 1996). Some members previously classified as *Pseudomonas* have been moved to other genera, including *Burkholderia*, *Stenotrophomonas*, and *Acidovorax* (Tables 1 and 2).

Risk Assessment of *Pseudomonas aeruginosa* in Water

Table 1 Classification of the genera *Pseudomonas*, *Burkholderia*, and *Stenotrophomonas*

Species	16s rRNA group	Characteristics
Pseudomonas aeruginosa *P. fluorescens* *P. putida* *P. stutzeri*	Gamma	Most produce water-soluble, yellow-green fluorescent pigments; do not easily form poly-β-hydroxybutyrate
Burkholderia cepacia *B. pseudomallei* *B. mallei*	Beta	No fluorescent pigments; form poly-β-hydorxybutyrate
Stenotrophomonas maltophilia	Gamma	Requires methionine; does not use NO_3^- as a nitrogen source

Kersters et al. (1996) and Gilligan et al. (2003)

Table 2 Old and new genera names for *Pseudomonas*

Old name	New name
P. cepacia	*Burkholderia cepacia*
P. maltophilia	*Stenotrophomonas maltophilia*
P. pseudomallei	*B. pseudomallei*
P. gladioli	*B. gladioli*
P. mallei	*B. mallei*
P. pickettii	*B. pickettii*
P. acidovorans	*Comamonas acidovorans*
P. testosteroni	*C. testosteroni*
P. delafeldii	*Acidovorax delafeldii*
P. facilis	*A. facilis*
P. temerans	*A. temerans*
P. paucimobilis	*Sphingomonas paucimobilis*
P. mesophilica	*Methylobacterium extorquens*
P. luteola	*Chryseomonas luteola*
Flavimonas oryzihabitans	*P. oryzihabitans*

The genus *Pseudomonas* consists of straight or slightly Gram-negative, curved (not helical) rods, 0.5–1.0 × 1.5–5.0 µm (Anon 1994). Many species accumulate poly-β-hydroxybutyrate as a carbon reserve material. Some members produce water soluble pigments. Motility occurs by one or several polar flagella; they are rarely nonmotile. In some species, lateral flagella of shorter wavelength may also be formed. They are aerobic, having a strictly respiratory type of Metabolism, with oxygen as the terminal electron acceptor; in some cases nitrate can be used as an alternate electron acceptor, and when present allows growth to occur anaerobically. Most, if not all, species fail to grow under acidic conditions of pH 4.5 or less.

3 Medical Importance

Most *Pseudomonas* infections result from contact with, or the use of, contaminated fluids (water, blood, and disinfectants) in the medical setting. Except for *P. aeruginosa*, most infections induced by pseudomonads only occur in individuals who have experienced trauma (i.e., wounds or surgery), or are immunocompromised.

Several *Pseudomonas* species have been associated with disease in humans, including *Pseudomonas fluorescens*, *Pseudomonas stutzeri*, and *P. aeruginosa*. *P. fluorescens* is able to grow at 4°C and, in rare cases, infection has been caused by contaminated infused blood. *P. stutzeri* is an unusual cause of human infection but has been reported to cause bacteremia in immunocompromised patients, patients with previous trauma, and pneumonia in alcoholics (Noble and Overman 1994). A cause of community-acquired vertebral osteomyelitis in a previously healthy patient has also been reported (Reisler and Blumberg 1999). *Pseudomonas puckettii* has been associated with sterile water used for injection (Roberts et al. 1990).

Of all the pseudomonads, *P. aeruginosa* is the most significant as a human pathogen in both normal and compromised individuals (Fick 1992). In addition, *P. aeruginosa* is the most important human pathogen in the genus *Pseudomonas*, *Stenotrophomonas*, or *Burkholderia* with respect to both numbers and types of infections, and associated morbidity and mortality. *Pseudomonas* species other than *P. aeruginosa* infrequently cause infection, in large part, because of their lack of virulence. Because of the low virulence of these other pseudomonads, the infections they cause are often iatrogenic and are associated with the administration of contaminated solutions, medicines, and blood products or the presence of indwelling catheters (Gilligan 1995; Lermi and Cunha 1998). In healthy individuals, they are responsible for eye, ear, and skin infections from contaminated water.

P. aeruginosa is the best studied of the pseudomonads because of its importance as a nosocomial agent; it is ranked as the fifth most frequent cause of nosocomial infection (Dembry et al. 1998). Other members of the genus rarely cause such infections and are not ranked. *P. aeruginosa* caused 0.56% of cerebrospinal fluid (CSF) infections, while the other unranked pseudomonads caused <0.06% of such cases. *P. aeruginosa* was also documented as the cause of 3.3% of bacteremia in patients, while other pseudomonads were not ranked and caused <0.005% of bacteremic cases (Farmer 1995). *Pseudomonas* bacteremia primarily affects immunocompromised patients who have significant underlying disease (Aksamit 1992). *Pseudomonas* is also the leading cause of burn wound infections (Gregory and Schaffner 1987). Furthermore, cystic fibrosis (CF) lung disease is unique in its association with chronic *P. aeruginosa* colonization. *P. aeruginosa* infects the respiratory tracts of as many as 60–87% of CF patients, and such infections may last for decades (Fick 1992). *P. aeruginosa* has become the predominate infective organism in CF patients, because these patients are surviving longer and are living longer (McCubbin and Fick 1992).

P. aeruginosa is also an important cause of nosocomial pneumonia (Bryan and Reynolds 1984). It produces a high mortality rate and is the only lethal *Pseudomonas* species specifically identified (Table 3). *P. aeruginosa* was the only pseudomonad identified by Torres et al. (1991) when he addressed community-acquired pneumonia. Many cases of nosocomial pneumonia occur in ventilator-assisted patients. The most common pneumonia-causing agent in these cases was *P. aeruginosa*; no other species of the genus received a ranking (Fagon et al. 1989). Eighty-seven percent of patients with pneumonia caused by *P. aeruginosa* or *Acinetobacter* spp. died,

Table 3 Bacterial causes of nosocomial pneumonia

Bacterium	Percent mortality attributed to pneumonia
P. aeruginosa	72
Klebsiella-Enterobacter-Serratia	40
S. aureus	33
E. coli	31
S. pneumoniae	29
Other aerobic gram-negative bacteria	25
Other streptococci	6

Bryan and Reynolds (1984)

Table 4 *Pseudomonas aeruginosa* infections associated with normal healthy individuals

Site of infection	Illness	Infection associated with
Eye	Keratitis	Contact lenses
Ear	External otitis	Swimming
Skin	Folliculitis	Whirlpools, hot tubs, swimming

whereas only 55% of patients with pneumonias caused by other organisms died. A study by Torres et al. (1990) also confirmed that *P. aeruginosa* is the only important pseudomonad that causes pneumonia in mechanically ventilated patients.

P. aeruginosa is also a leading cause of sepsis in hospitalized patients and is the fifth most frequent organism isolated from blood cultures in cases of septicemia (Trautmann et al. 2005). In an evaluation of 612 episodes of gram-negative sepsis over a 10-yr period, Kreger et al. (1980) found that the predominant etiological agents were (in descending order) *Escherichia coli*, *Klebsiella–Enterobacter–Serratia* species, *P. aeruginosa*, *Proteus* and *Providencia* spp., and *Bacteroides* spp. Of a total of 60 cases of *Pseudomonas* sepsis, 54 of the isolates were identified as *P. aeruginosa*. Other pseudomonads were not identified.

4 Health Effects in Humans

According to Hardalo and Edberg (1997), the range of infections caused by *P. aeruginosa* is narrow and related to specific changes in the defense and immune status of the host. They claim (multiple times) that the bacterium does not attack normal tissue. In many cases, this is true. However, *P. aeruginosa* is a well-established cause of ear and/or skin infections in healthy noncompromised hosts (Table 4). It may be argued that extended water-to-skin contact damages tissue by removing fatty acids, oils, and keratin layers. This type of damage is unavoidable, as are minor tiny scratches caused by the wearing of contact lenses which can predispose a person to infection. The type of infections caused by *P. aeruginosa* is actually quite large, and this bacterium is a leading cause of illness in compromised hosts (Table 5).

Table 5 *Pseudomonas aeruginosa*-induced nosocomial infections associated with immunocompromised individuals

Septicemia (blood infection)
Endocarditis (heart disease)
Osteomyelitis (bone infection)
Urinary tract
Gastrointestinal disorders
Pneumonia
Respiratory tract
Meningitis (nervous system)
Burn patients (infection of the skin)

Hardalo and Edberg (1997) also reported that we ingest millions of *P. aeruginosa* in our food daily. However, the references cited to support this point are articles that address the presence of *P. aeruginosa* in hospital foods. There is little information that addresses the presence of *P. aeruginosa* in household foods or in other nonhospital environments.

4.1 Nosocomial Disease

The majority of serious *P. aeruginosa* infections are hospital acquired (Bodey et al. 1983). *P. aeruginosa* is primarily a nosocomial pathogen and is responsible for almost 10% of nosocomial infections (Dembry et al. 1998). The percent of nosocomial diseases caused by *P. aeruginosa* has been reported as follows: urinary tract infection, 11%; surgical site infections, 8%; pneumonia, 16%; and septicemia, 3–4% (Dembry et al. 1998; Trautmann et al. 2005). This organism is a major cause of nosocomial pneumonia, and is the third most common cause of nosocomial urinary tract infections and surgical wound infections. Patients at the greatest risk of infection are burn patients, cystic fibrosis patients, individuals infected with HIV, elderly persons, and premature infants (Botzenhart and Doring 1993).

The incidence of colonization by *P. aeruginosa* in healthy persons, or those entering hospitals is relatively low. Representative site-specific colonization rates are skin (0–2%), nasal mucosa (0–3.3%), throat (0–6.6%), and stool (2.6–24%; Pollack 1995). During hospitalization, carriage rates may be greatly increased, especially in burn patients, in the lower respiratory tract of patients undergoing mechanical ventilation, in the gastrointestinal tract of patients receiving chemotherapy, or at virtually any site in persons treated with antibiotics. In each case, colonization rates may exceed 50%, and colonization often presages invasive infection (Tancrede and Andremont 1985). Although colonization by *P. aeruginosa* frequently precedes overt infection, the original source of the organism and the mode of transmission are often unclear.

4.2 Septicemia

Fewer than 100 cases of *Pseudomonas* bacteremia was reported prior to 1950. With the incorporation of potent immunosuppressive regimens into clinical practice to treat neoplastic and inflammatory diseases in the 1950s and 1960s, dramatic increases occurred (Aksamit 1992). Only one large published study included both Gram-positive and Gram-negative causes of bacteremic infections (McGowan et al. 1975). This study, which addressed a 12-yr span at Boston City hospital, showed that *P. aeruginosa* caused 3.6% of the cases and was associated with the highest fatality rate. When only Gram-negative bacteria were ranked, *Pseudomonas* was implicated with a frequency of 10.9–12.4% (Greenman et al. 1991; Ziegler et al. 1991).

P. aeruginosa causes septicemia primarily in immunocompromised patients. Predisposing conditions include hematologic malignancies, immunoglobulin deficiency, neutropenia, diabetes mellitus, organ transplantation, severe burns, and AIDS (acquired immunodeficiency syndrome; Pollack 1995). Other predisposing conditions include chemotherapy, steroid administration, antibiotic therapy, intravenous (IV) lines, urinary catheterization, surgery, trauma, and premature birth. Mortality remains high despite advances in therapy. The fact that *P. aeruginosa* may cause high mortality, even in patients without severe underlying disease, emphasizes its pathogenic potential in the bloodstream.

4.3 Endocarditis

Endocarditis is a microbial infection of the heart valves or of the endocardium. *P. aeruginosa* cases are sometimes associated with burn patients, intravenous drug use, hemodialysis, and open heart surgery (Molina et al. 1991). Such cases of endocarditis were very rare prior to the last two decades. In fact, before 1973 it accounted for <0.25% of all cases seen at the New York Hospital during a 30-yr period (Saroff et al. 1973). However, since then, there has been a significant increase from illicit use of drugs such as heroin (Shekar et al. 1985). More than 90% of all reported *P. aeruginosa* endocarditis afflicts IV drug users (Pollack 1995).

4.4 Osteomyelitis

It is unusual for osteomyelitis to be caused by *P. aeruginosa*, except as a complication of septicemia or puncture wounds. Predisposing factors such as surgery, compound fractures, penetrating wounds, and drug addictions increase the probability of *P. aeruginosa* osteomyelitis (Bodey et al. 1983). Reports have shown an increase in the frequency of such osteomyelitis cases primarily because of an increase in drug users (Wiesseman et al. 1973; Salahuddin et al. 1973).

P. aeruginosa is the most common pathogen implicated in osteochondritis following puncture wounds of the foot. It was originally described in children but may also occur in adults. The majority of children who contacted this infection were wearing tennis shoes at the time of injury; *P. aeruginosa* has been isolated from the soles of the shoes in some cases (Jacobs et al. 1989). This infection involves the small joints and bones of the foot.

4.5 Pneumonia

The overall attack rate of pneumonia in the US is 12–15 cases per 1,000 persons per yr (Garibaldi 1985; Marrie 1994) resulting in ca. 3,957,000 cases annually. Pneumonia is the sixth leading cause of death in the US, and has an estimated annual cost of $23 billion (Marrie 1994).

P. aeruginosa was found to be the etiological agent for up to 5.0% of cases and is associated with a high mortality (Garibaldi 1985; Torres et al. 1991). *Pseudomonas* pneumonia is notoriously difficult to treat (Bodey et al. 1985), and reports suggest that *P. aeruginosa* pneumonia is on the increase (Bodey et al. 1983). This infection usually occurs in patients with cystic fibrosis, hematologic malignancies, diabetes, chronic lung or heart disease, and is predominantly nosocomial. Exposure to the hospital environment (particularly in an intensive care setting), the use of respiratory inhalation equipment, and prior antibiotic therapy increase the probability of contacting *P. aeruginosa* pneumonia. Celis et al. (1988) reported the mortality rate associated with *P. aeruginosa* pneumonia to be as high as 70%, and causes up to 90% of deaths in patients with cystic fibrosis.

Aspiration of *P. aeruginosa* may also produce pneumonia. Despite rigorous cleaning of ventilator equipment and frequent changes of disposable tubing, direct aerosolization of the bacterium may occur in patients receiving respirator therapy. Nebulizers introduce bacteria directly into the lower respiratory tract.

Bacteremic pneumonia occurs primarily in neutropenic patients after cancer chemotherapy (Pollack 1995). The disease begins with the inhalation of the bacteria followed by bloodstream invasion; effects include pulmonary lesions and sometimes lesions in other organs as well. Bacteremic *P. aeruginosa* is also a fulminant disease that is usually fatal within 3–4 d.

4.6 Urinary Tract Infections

P. aeruginosa-induced urinary tract infections (UTI) are usually nosocomial and iatrogenic. This bacterium is a leading cause of such nosocomial infections and may produce sepsis or other serious complications, depending upon the immune system status of the patient (Shigemura et al. 2006). It has also been reported that community-acquired *P. aeruginosa* UTIs in children are associated with peculiar clinical aspects when compared to those caused by other bacteria (Goldman et al. 2008).

4.7 Gastrointestinal Infections

P. aeruginosa-associated gastrointestinal disease may be underestimated because it is often unapparent, is difficult to distinguish from other causes, or is overshadowed by pathologic events outside the GI tract. This bacterium can produce disease in any portion of the GI tract, from the oropharynx to the rectum. Infection of the GI tract is primarily seen in immunocompromised individuals, especially young infants and neutropenic patients. A GI tract infection may lead to septicemia. Septicemia primarily occurs in hospitalized cancer patients, those taking broad-spectrum antibiotics, and patients who develop GI colonization with *P. aeruginosa*. Chemotherapy may result in neutropenia, while the colonized GI tract serves as a reservoir for septicemia. Although local signs of gastrointestinal involvement are sometimes apparent, infections are usually clinically inapparent and go unrecognized until giving rise to septicemia (Pollack 1995).

P. aeruginosa is thought to cause a severe necrotizing enterocolitis in young infants and in neutropenic cancer patients. This disease results in ulcerating lesions that begin in the bowel mucosa and extend into the submucosa (Amromin and Solomon 1962). The lesions are usually hemorrhagic and necrotic. In addition, anorectal infections are sometimes seen in neutropenic cancer patients, which may be accompanied by septicemia. These rectal abscesses can give rise to life-threatening infection and should be treated aggressively when they appear (Schimpff et al. 1972; Angel et al. 1991).

Typhilitis is one of the most serious GI infections caused by *P. aeruginosa* (Artenstein and Cross 1993), and is usually seen in leukemia patients (Bodey et al. 1983). It is a necrotizing enterocolitis characterized by localized necrotic and gangrenous lesions in the cecum, but may involve the entire colon. A high mortality rate has been associated with typhilitis, especially in those with septicemia (Artenstein and Cross 1993).

Only a few studies describe *P. aeruginosa*-caused GI infections in normal healthy individuals. Pollack (1995) suggested that a study by Ensign and Hunter (1946) left little doubt that *P. aeruginosa* caused an outbreak of diarrhea in school children. However, careful examination of the study shows that the authors described an outbreak of diarrhea in a neonatal hospital nursery caused by milk contaminated with *P. aeruginosa*; other diarrheal-causing pathogens were not considered.

In a study reported by Adlard et al. (1998), *P. aeruginosa* was the predominant organism isolated from the feces of 23 unrelated hospital outpatients investigated in the course of 1 yr for diarrhea lasting more than 1 wk. Patient histories were reviewed and virulence was determined by studying selected *in vitro* and *in vivo* fecal isolates. The patients had a mean age of 60 yr (range 3–90 yr), were receiving antibiotics, and/or had an underlying illness. Microbiological examination of the diarrheal feces included exams for parasites, rotavirus, *Campylobacter*, *Aeromonas*, *Salmonella*, *Shigella*, and *Clostridium difficile*. All but one of the patients were considered to be immunocompromised, either through age, underlying illness, past history, or from antimicrobial or other drug therapy. For ten patients (age range 74–90 yr), age was the only risk factor. Seven patients had underlying disease (including one each of hiatus hernia, gastritis, breast cancer, and pancreatitis). The authors believed that their findings suggest that *P. aeruginosa* can cause diarrhea, particularly in immunodeficient individuals.

Porco and Visconte (1995) described a case of gastroenteritis apparently caused by *P. aeruginosa* as a nosocomial infection in an immunocompromised patient. In this case, an 80-yr-old woman with diabetes mellitus and hypertension developed progressive renal insufficiency and underwent hemodialysis.

4.8 Burn Patients

P. aeruginosa is a leading cause of burn-wound infections in burn patients (Molina et al. 1991). According to Ollstein and McDonald (1980), a concentration of 10^5 organisms per gram of tissue is required to induce an invasive burn-wound sepsis. Surveys performed in one US burn institution have shown that from 1959 to 1983, 77% of the mortality in burn patients resulted from *P. aeruginosa* (McManus et al. 1985). The mortality of bacteremia in these patients did not change significantly over the 25-yr period, despite the development of new antibiotics. Although these infections have decreased in some institutions, they continue to be strongly associated with the morbidity and mortality of burn patients worldwide (Holder 1993).

The epidermal skin layers of burn patients have damage that produces necrotic tissue and increased moisture exudation that predisposes to *P. aeruginosa* infection (Solomon 1985). Although a Gram-positive flora predominate at the burn site in the immediate postburn period, it is soon replaced by Gram-negative bacteria, especially *P. aeruginosa*. It appears that hydrotherapy may promote the colonization of burn patients by multiresistant *P. aeruginosa* (Tredget et al. 1992). *Pseudomonas* rapidly proliferates in a burn wound, and becomes sufficiently invasive to produce septicemia. Burn-wound infections appear as black, dark brown, or violaceous discolorations. Degeneration of underlying granulation tissue occurs with edema and hemorrhagic necrosis. Pneumonia is also common in burn patients, particularly if there has been previous inhalation injury (Pollack 1995).

4.9 Skin Infections

P. aeruginosa folliculitis is the most common recognizable infectious disease occurring from use of whirlpools and hot tubs (Solomon 1985). However, several alternative diagnoses may be made unless the physician is aware of the history of the patient (Brett and du Vivier 1985). Folliculitis is a superficial or deep bacterial infection and irritation of the hair follicles. A rash may develop 8–48 hr after exposure to *P. aeruginosa*, which is characterized by the abrupt onset of a wide-spread itchy uriticarial papulas and deep pustules (Molina et al. 1991). Most skin infections are caused by serotype O:11 (Thomas et al. 1985). The rash may be accompanied by earache, malaise, fatigue, headache, mastitis, or a low-grade fever and axillary lymphadenopathy. In most cases, the rash heals in 7–10 d without scarring. The incubation period varies from 8 hr to 5 d, with a mean of 48 hr (Gregory and Schaffner 1987). Two distinct types of *P. aeruginosa* folliculitis have been reported: A sudden unmanageable exacerbation of acne that may result from a superimposed *P. aeruginosa* infection, and a more common, less severe type of folliculitis. The latter is usually associated with the use of public hot tubs, whirlpools, and swimming pools (Molina et al. 1991).

Skin dryness is normally a defense against infectious agents but this defense is altered by immersion in water. After 20 min of contact with water, the water content of skin may increase by 55–70%. With prolonged immersion, the water content of the stratum corneum may increase 25–30 times. An intrinsic host factor for *P. aeruginosa* skin infections is the age of the host. In many reported outbreaks of whirlpool-related disease, cases occurred in adolescents and young adults. This association may be related to different patterns of use of hot tubs by persons in different age groups, but may be also be skin-related, because the skin does undergo significant changes with growth and aging. Apocrine skin glands become active at puberty and the quantity and quality of the sebaceous gland secretions change markedly between childhood and adolescence (Solomon 1985). However, Jacobson (1985) reported no relationship between age and risk of skin infection, and immunocompromised persons have not been identified as predisposed to skin infections. Although this illness may be treated by a physician, most patients do not miss work or school. Furthermore, *P. aeruginosa* is known to produce proteases. If proteases are produced in hot tub water, they may contribute to the likelihood of colonization by damaging the skin surface by increasing skin permeability and water content. Despite subsequent drying of the skin surface, *P. aeruginosa* may colonize hair follicles and produce proteolytic enzymes and exotoxins that foment inflammatory reactions (Solomon 1985).

P. aeruginosa may also produce green nail syndrome, especially if the nail is frequently exposed to water. An example is a dishwasher afflicted with recurrent onycholysis; such individuals have persistent greenish pigmentation of the thumb nail. After removing the onycholytic portion of the nail and soaking in dilute (1%) acetic acid, the problem can be resolved (Agger and Mardan 1995).

Gram-negative bacteria can cause skin infections involving the toe web space due to the presence of moisture; *P. aeruginosa* is the bacterium most frequently associated with these infections (Molina et al. 1991). Toe web infection is often found among military personnel, in which the macerated scaly webs reveal a greenish fluorescence under a Woods lamp. Tinea pedis is a common antecedent to toe web infection (Molina et al. 1991).

4.10 Ear Infections

P. aeruginosa is not a normal component of ear flora (Molina et al. 1991) and is only found in 1–2% of healthy human ears (Hall et al. 1968); however, *P. aeruginosa* accounts for 70% of cases of external otitis (swimmer's ear). Swimming during the preceding week is strongly associated with such ear infections. Swimmers with otitis externa were more likely to have swum longer, more frequently, and with more frequent submersion of their heads than swimmers without otitis externa (Springer and Shapiro 1985). Swimming in fresh water or pool water is more likely to be associated with ear infections than swimming in ocean water (Calderon and Mood 1982; Springer and Shapiro 1985). However, there is a strong association between skin diving in seawater and colonization of the ear and the upper respiratory tract (Losonsky et al. 1994). Individuals younger than 18 yr are also more likely to develop swimmer's ear (Seyfried and Cook 1984). This could result from more swimming activity by younger persons (Calderon and Mood 1982). External otitis, caused by this bacterium, is associated with the presence of *P. aeruginosa* in inadequately chlorinated whirlpools; iatrogenic causes include trauma induced by forceful ear irrigation (Molina et al. 1991).

As measures of water quality, numbers of fecal coliforms, enterococci, *P. aeruginosa*, *Staphylococcus*, or total plate counts do not necessarily correlate with incidence of ear infections. One possible explanation is bather pollution; otitis externa may occur despite good bacterial recreational water quality. Large numbers of bacteria are shed by humans when swimming. *P. aeruginosa*, although only shed in small numbers, could transiently exist long enough in water to enter the ear (Calderon and Mood 1982). Another possibility these authors did not consider is that injured or nutrient-stressed *Pseudomonas* may not grow well on selective agar without prior resuscitation.

Ear infection is facilitated by ear injury, maceration, humidity, or ear moistness (Dinapoli and Thomas 1971). Symptoms of ear infection include nocturnal pain, and headache, which is remarkable for assessing the severity of the infection. Tenderness and swelling about the periauricular area is common, and purulent drainage is sometimes noted. Early damage to the cranial nerve is reversible, but if allowed to progress, can be permanent (Molina et al. 1991).

If treatment fails, external otitis may become invasive, especially in elderly diabetic patients (Bodey et al. 1983; Molina et al. 1991). Elderly patients afflicted with diabetes mellitus have a diminished cutaneous barrier against infection because of diabetes-induced microangiopathy (Molina et al. 1991). Aging and diabetes mellitus are also associated with reduced immune function. If ear infections become invasive it is called malignant otitis externa, and may progress to basilar skull osteomyelitis and/or meningitis (Gilligan 1995).

4.11 Eye Infections

P. aeruginosa keratitis is among the most common ocular infections and can potentially lead to loss of the eye (Bodey et al. 1983). About 50% of corneal ulcerations are caused by *P. aeruginosa* (Molina et al. 1991). The incidence of ulcerative keratitis caused by *P. aeruginosa* is estimated to be 1 in 500 for persons using extend-wear contact lenses. *P. aeruginosa* is the most common organism associated with contact lens-related keratitis (Fletcher et al. 1993). Infections of the cornea usually begin with some form of trauma, often minor, which causes a break in the epithelial surface that invites *P. aeruginosa* invasion (Ramphal et al. 1981; Fletcher et al. 1993). Contaminated contact lens solutions have also been implicated as sources of infection (Gilligan 1995). Multiple predisposing factors caused by wearing soft contact lenses include epithelial hypoxia, mechanical abrasion, preservative hypersensitivity, and giant papillary conjunctivitis. Although several types of contact lenses predispose to infection, those wearing soft contact lenses appear to be most susceptible. In addition, burn patients (thermal damage to the cornea), premature infants, and patients with predisposing ocular conditions are especially susceptible to this bacterium (Pollack 1995). This disease is characterized by a sudden onset, rapid circumferential spread, and sloughing of the cornea, with the potential for early perforation and poor visual outcome (Molina et al. 1991).

4.12 Meningitis

The first case of *P. aeruginosa* meningitis was reported in 1983 (Bodey et al. 1983). This bacterium accounted for 9% of central nervous system infections in cancer patients over a 16-yr period (Chernick et al. 1973). Individuals with the greatest risk for *P. aeruginosa* meningitis are those with CSF shunts or reservoirs and those with tumors of the head and neck (Fong and Tomkins 1985). The disease onset may be acute or fulminant with fever, headache, confusion, and coma. In some cases, the onset is gradual. This latter case is common in immunosuppressed or cancer patients, and in those whose meningitis results from neurosurgery or extension from a contiguous site of chronic infection (Berk et al. 1987).

5 Special Populations

5.1 Children

A survey disclosed that only <0.3 to 1.0% of 310 healthy babies carried *P. aeruginosa* in the nasopharyngeal tract. Although *P. aeruginosa* is not common, babies as young as 1 month carried the bacterium (Faden 1998). Although healthy children exposed to *P. aeruginosa* usually resist disease, premature infants are more susceptible; this may result in nursery outbreaks of septicemia, meningitis, and pneumonia (Bobo et al. 1973). *P. aeruginosa* is also a serious complicating pathogen in newborns that have perforated necrotizing enterocolitis (Stone et al. 1979). In this study, bacterial cultures were taken from blood, the peritoneal cavity at surgery, and from any subsequent wound and/or intraperitoneal infection. No significant differences in presence of anaerobes in peritoneal fluid were noted between fatal and nonfatal.

However, peritonitis was invariably fatal when *P. aeruginosa* was present (seen in 16.4% of these patients) or *P. aeruginosa* septicemia existed.

Cystic fibrosis (CF) is a genetic disease that begins in childhood, in which abnormal respiratory secretions impede normal pulmonary clearance functions. In afflicted children, the incidence of lower respiratory tract infection with mucoid strains of *P. aeruginosa* ranges from 21% (< 1 yr of age) to more than 80% (26 yr or older). Cystic fibrosis patients are living longer than in the past (Pollack 1995). It is unclear whether mucus plugging precedes infection or vice versa. Initially, patients may have frequent upper respiratory tract infections with a lingering cough after each episode. There may be recurrent bouts of pneumonia. Most patients eventually develop a chronic productive cough, decrease in appetite, weight loss, and diminished activity. Physical signs include evidence of undernutrition, cyanosis, wheezing, moist rales, abdominal distension, and clubbing of fingers and toes. Mucus plugging of the dilated bronchi is often observed (Pollack 1995).

An unusual mucoid phenotype of *P. aeruginosa* is associated with chronic lung infection in 70–80% of adolescent and adult CF patients. The unusual lung environment of the CF patient is believed to switch on a cluster of genes that encode for production of large amounts of alginate, a polysaccharide polymer. Alginate appears to be antiphagocytic and appears to induce an immune response. This immune response is partially responsible for the lung damage in these patients (Gilligan 1995). There is no cure for *P. aeruginosa* pulmonary infections in CF patients, but antibiotic therapy has improved survival. Unfortunately, antibiotic use in these patients is frequently associated with the emergence of antibiotic-resistant strains (Pollack 1995).

The mode of transmission of *P. aeruginosa* to CF patients is often not clear. Clinic exposures and/or social interactions with other CF patients are thought to facilitate the spread of *P. aeruginosa* between CF patients (Farrell et al. 1997). CF patients may also acquire *P. aeruginosa* outside of hospitals (Doring et al. 1989; Remington and Schimpff 1981). The study by Doring et al. (1989) showed that 67% of CF patients with chronic *P. aeruginosa* lung infections carried the bacterium in the GI tract. Strains of *P. aeruginosa* in the sputa and in stool cultures of these patients appeared to be identical. However, no differences were seen in the health of patients with *P. aeruginosa*-negative stool samples and patients with positive stool samples. Toilets in households of the *P. aeruginosa*-infected CF patients were significantly more contaminated with *P. aeruginosa* (42%) than toilets in households of noninfected patients (20%). This may be significant, because one study showed that bacterial-laden aerosols may be created by flushing the toilet (Rusin et al. 1998). In one study (Friend and Newsom 1986), two young patients with CF simultaneously acquired infection with serotype O:10 following swimming lessons in a hydrotherapy pool. Subsequent examination of the water revealed *P. aeruginosa* O:10 in the pool water.

As mentioned, individuals younger than 18 yr are more likely to develop swimmer's ear than are older persons (Seyfried and Cook 1984). This may result from more swimming activity by younger persons (Calderon and Mood 1982). Also discussed previously, is the tendency for younger people to acquire *P. aeruginosa* skin infections more easily after exposure to whirlpools or swimming pools than do older people. This association may result from different patterns of hot-tub use by persons in different age groups; the skin is known to undergo significant changes with growth and aging. Apocrine skin glands become active at puberty and the quantity and quality of the sebaceous glands secretions change markedly between childhood and adolescence (Solomon 1985).

5.2 AIDS Patients

There is a growing concern because recent studies have shown an increased incidence of *P. aeruginosa* infections among AIDS patients (Mendelson et al. 1994). These authors evaluated 27 episodes of *P. aeruginosa* septicemia in 21 AIDS patients over a 6-yr period. Of 21 primary cases, 12 were community acquired, 8 were nosocomial, and one was acquired in a nursing home. Sources of the septicemia included lung infections, indwelling vascular catheters, the upper respiratory tract, and unknown sources. Mortality was 52.6%, which occurred despite treatment with an appropriate antibiotic. In another study of 336 AIDS patients, only two patients developed septicemia through contact with *P. aeruginosa* (Whimbey et al. 1986). In a limited study of 59 patients by Witt et al. (1987), 31 had AIDS-related-complex and 28 had AIDS. Of these, 81% of infections were community acquired and 19% were nosocomial. One of the 59 patients had two separate episodes of nosocomial pneumonia caused by *P. aeruginosa*, and accompanied by bacteremia. A study by Kiehn (1989) showed that most bacteremia, in AIDS patients, was caused by *Mycobacterium avium*. Among other causes, *P. aeruginosa* ranked seventh and caused septicemia in 3 of the 336 patients. In a similar study of septicemia in AIDS patients (Eng et al. 1986), *P. aeruginosa* received a similar ranking of eighth among the etiological agents. However, these rankings are similar to those found in general septicemia studies in which *P. aeruginosa* has been ranked from number 3 to number 11 as the most frequent cause (McGowan et al. 1975; Kreger et al. 1980). In addition, Eng et al. (1986) discovered that the bacteremia caused by *P. aeruginosa* was less frequently found in patients with AIDS, than in patients without AIDS.

P. aeruginosa is increasingly reported as a respiratory pathogen in patients with advanced HIV disease (Traill et al. 1996). In one study of 37 HIV-positive patients, 35 were men with a mean age of 36 yr. Two patients were African heterosexuals. All 37 patients exhibited acute *P. aeruginosa* bronchopulmonary disease. Infection was community acquired in 62% of the patients and nosocomial in 38%. Of the 37 patients, three had respiratory copathogens including *Hemophilus influenza*, *Streptococcus pneumoniae*, and *Mycobacterium avium*-intracellulare (one patient each). Five of the patients had pulmonary Kaposi sarcoma. Pulmonary infection in patients such as these appears to be a feature of advanced HIV disease. Recent reports of *P. aeruginosa* respiratory infection in HIV patients have observed a higher percentage of community-acquired cases as noted by Traill et al. (1996). Respiratory infections in AIDS patients with *P. aeruginosa* are often a chronic condition. Multiple isolations of the same strain of *P. aeruginosa* may be made (Asboe et al. 1998). It may be difficult to eradicate the bacterium from these patients' respiratory tracts, despite appropriate therapy. Another study reported that severe cases of *P. aeruginosa* infection occurs in HIV patients including pneumonia, sepsis empyema, malignant otitis externa, UTI, keratitis, cellulitis, and chest wall abscesses (Kielhofner et al. 1992).

HIV-infected children are also at increased risk for pseudomonal infection (Flores et al. 1993). Six episodes of *P. aeruginosa* septicemia in five children with AIDS were reviewed. The ages of the patients ranged from 5.75 mon to 11 yr. In addition to existence in blood, *P. aeruginosa* – positive sites included the skin, peritoneal fluid, the ear, the cerebral spinal fluid, and sputum. In four of the six patients, the lungs were infected. All but one patient exhibited hypotension and two of the patients died. The authors neither gave the time period over which the study extended, nor ranked *P. aeruginosa* against other agents capable of causing septicemia in the study population.

5.3 Granulocytopenic Patients

Some organisms, such as *P. aeruginosa*, are particularly virulent in granulocytopenic patients. Septicemia may develop in 40–70% of patients who are colonized with *P. aeruginosa* (Remington and Schimpff 1981). Patients with acute leukemia (Bodey et al. 1985) or non-Hodgkin's lymphoma (Bishop et al. 1981) are susceptible to septicemia by *P. aeruginosa*. In the latter study, *P. aeruginosa* was the third leading cause of infection in the patients, who suffered a 28% fatality rate. Most of these patients are neutropenic at the onset of infection. Most of these infections appeared to be nosocomial. Factors associated with a poor prognosis included shock and pneumonia.

5.4 Cancer Patients

The availability of new, more effective, antipseudomonal antibiotics has profoundly affected the course of *P. aeruginosa* septicemia in cancer patients. Formerly, only 24% of such patients were cured, whereas, a study by Bodey et al. in 1985 showed a 66% cure rate. Improperly treated *Pseudomonas* bacteremia can rapidly be fatal. In the study by Bodey et al. (1985), the mortality during the first 24 hr was 26% among those who received inappropriate antibiotics, compared with 5% among those appropriately treated.

6 Occurrence and Survival of *P. aeruginosa*

Water is the natural reservoir for *P. aeruginosa* (Molina et al. 1991). Some authors consider the presence of *P. aeruginosa* to be an indicator of contamination from surface runoff, domestic and agricultural effluents, or human fecal matter (Warburton et al. 1994). The major source of *P. aeruginosa* in surface water is thought, by some authors, to be domestic sewage (de Vicente et al. 1988), and *P. aeruginosa* is found in 90% of sewage samples (Geldreich 1996). Its concentration in surface waters receiving waste and stormwater discharges ranges from 1 to 10,000 cells/100 mL; in stormwater, concentrations are 1,000–100,000 mL^{-1} (Geldreich 1996). However, the intestinal carriage rate for *P. aeruginosa* in humans is low, suggesting that its presence in water does not necessarily result from sewage contamination. Other sources of the organism are thought to be water leaching from agricultural soils, barnyard drainage, and urban run-off.

Although considered by some to be environmentally ubiquitous, others believe the presence of *P. aeruginosa* is linked directly to human activities (Relai and Rosati 1994). These latter authors reported up to 2,400 *P. aeruginosa*/100 mL in surface waters heavily contaminated by fecal discharge, while it was absent in less polluted water. They also found that *P. aeruginosa* could always be isolated when sampled from distribution plants (reservoirs and pipes), but not from wells or springs. Isolates from environmental waters tend to be less resistant to antibiotics than those from clinical specimens (Relai and Rosati 1994).

6.1 Drinking Water

P. aeruginosa is considered to be the most important pseudomonad in drinking water (Geldreich 1996). Allen and Geldreich (1975) found that 3% of drinking water samples contained *P. aeruginosa*, with counts ranging between 1 and 2,300 organisms per milliliter. Geldreich (1996) later expressed concern for the presence of *P. aeruginosa* in the water supply because of its potential to increase exposure to senior citizens, AIDS patients, and infants. He believed ingestion was the route responsible for transmitting this organism. He suggested that amplification may occur in static water areas and in sediment-laden water pipes during warm-water periods. Gambassini et al. (1990) found *P. aeruginosa* in only 1 of 66 drinking water samples; the single detection was from a sample at the terminal end of a distribution system. In addition, carbon filter point-of-use (POU) devices may become colonized with *P. aeruginosa*. Therefore, Geldreich (1996) has suggested that these devices should not be used on untreated water supplies of questionable quality, unless they are fitted with an associated microbial barrier.

A survey performed in southern Greece showed that *P. aeruginosa* occurred in 9% of tap water samples at a mean concentration of seven colony-forming units (CFU)/100 mL, and in 18.8% of bottled water samples at a mean concentration of 1,000 CFU/100 mL (Papapetropoulou et al. 1994). These authors noted that the *P. aeruginosa* isolates were resistant to certain antibiotics such as tetracycline, streptomycin, and naladixic acid. A synopsis of the occurrence of *P. aeruginosa* in potable water is shown in Table 6.

It would appear that the major exposure of *P. aeruginosa* via tap water results from colonization of the tap fixtures (i.e., faucets, showerheads, sinks, etc.) and introduction into the water when these devices are used. In one study in a children's

Table 6 Occurrence of *P. aeruginosa* in potable water

Type of water	Percentage of positive samples or concentration found	Reference
Tap water	<2.0	Levesque et al. (1994)
Tap water	9, 7 CFU/100 mL	Papapetropoulou et al. (1994)
Tap water	<10 (none detected)	Payment et al. (1988)
Tap water	0–0.25 CFU/L	Payment et al. (1988)
Tap water	0.01–0.6	Clark et al. (1982)
Tap water	3, 1–2,300 mL^{-1}	Allen and Geldreich (1975)
Tap water	24.1, 0.01–0.38 mL^{-1}	Reitler and Seligmann (1957)
Treated drinking water	1.5	Gambassini et al. (1990)
Mineral water	8.3	Manaia et al. (1990)
Water cooler water	<2.0	Levesque et al. (1994)
Water vending machine	16.1, 0.002–0.16 mL^{-1}	Chaidez et al. (1999)
Bottled water	18.8, 1,000 CFU/100 mL	Papapetropoulou et al. (1994)
Bottled water	0.5	Warburton et al. (1992)
Bottled water	<0.4 (0 of 256)	Duquino and Rosenberg (1986)

CFU Colony-forming units

hospital in Germany, 81% of all sinks were contaminated with *P. aeruginosa* (Doring et al. 1991). Of 484 tap water samples collected from 39 faucets in a medical intensive care unit, more than 11% contained *P. aeruginosa* (Rogues et al. 2007). Other exposures via tap water may result from the growth of *P. aeruginosa* in vending machines (Schillinger and Du Vall Knorr 2004), drinking water dispensers (coolers) (Baumgartner and Grand 2006), and water softeners (Hambsch et al. 2004).

6.2 Recreational Waters – Hot Tubs, Whirlpools, and Swimming Pools

The presence of *P. aeruginosa* in recreational waters has been associated with outer ear and skin infections. Ratnam et al. (1986) described outbreaks of *Pseudomonas* folliculitis associated with the use of whirlpools and swimming pools for the years 1972–1985 in the USA, Canada, and England. The authors sampled one whirlpool on two consecutive days and reported the following concentrations: 900 CFU/100 mL on the first day, and 220,000–340,000 CFU/100 mL on the second day. Whirlpools are normally operated at 39–40°C. Bacterial populations may rise rapidly if disinfectant concentrations fall below the recommended levels of 3.0 ppm chlorine or 6.0 ppm bromine (Price and Ahearn 1988). Hollyoak et al. (1995) reported that all 17 whirlbaths in 16 sampled nursing homes harbored *P. aeruginosa*, despite the fact that many were routinely disinfected. Water in a whirlpool bath, unlike a spa pool, is not filtered or chemically treated, although the bath is drained and cleaned after use by each bather. However, it was found that only one of 253 nursing home residents, who used whirlpool baths, had a known *P. aeruginosa* infection. No epidemiological typing of the infectious agent was done. Hollyoak et al. (1994) reported high levels of *P. aeruginosa* contamination, despite routine use of disinfectants such as quaternary ammonium compounds, phenolics, and hypochlorite. P. aeruginosa was not found in the water from the hot or cold taps.

Price and Ahearn (1988) recovered P. aeruginosa from all commercial whirlpools and residential units tested. P. aeruginosa was also recovered from aerosol samples from both residential and commercial whirlpools. P. aeruginosa populations levels (10^4–10^6 cells/mL) were found in the water within 96 hr after use of disinfectant addition (1-bromo-3-chloro-5,5 dimethylhydantoin or dichloroisocyanurate) ceased. At a whirlpool water temperature of 39–41°C, with the jets in operation, chlorine levels decreased from >3 ppm to <1 ppm within 1 hr.

Levels of 10–50 CFU/mL of P. aeruginosa have been found in hydrotherapy pools, despite prechlorination levels of 0.2–3.0 ppm and postchlorination levels of 1.25–4.0 ppm free chlorine (Aspinall and Graham 1989). The off-line pump water yielded >10^5 CFU/mL, whereas the on-line pump yielded no Pseudomonas. Hose pipes are also a source of P. aeruginosa contamination in whirlpools. A summary of P. aeruginosa in whirlpools and swimming pools not linked to disease is shown in Table 7. Table 8 provides a synopsis of the occurrence of P. aeruginosa in whirlpools and swimming pools associated with outbreaks of disease.

6.3 Surface Waters

P. aeruginosa is sometimes found in rivers; higher numbers are found near sites of urban runoff. Concentrations at a site near a town of 20,000 ranged from 10^2 to 10^3/100 mL, whereas concentrations further downriver ranged from <1 to 10^2/100 mL (Alonso et al. 1989). The proximity of the lake sampling sites to sewage treatment plants may influence the rate of recovery of P. aeruginosa. At locations upstream of treatment plants, the range was 2–33 CFU/100 mL, whereas 350/100 mL was isolated from a site downstream at an emergency raw-sewage overflow pipe (Seyfried and Cook 1984).

Table 7 Presence of P. aeruginosa in whirlpool and swimming pool waters

Type of water	Concentration of P. aeruginosa (100 mL)	Free chlorine concentration (mg/L)	Reference
Swimming pool	0.7–18.7	0.02–0.42	Falcao et al. (1993)
Whirlpool	<2–2,400	0.4–3.0	Kush and Hoadley (1980)
Whirlpool	<2–33	3.0	Kush and Hoadley (1980)
Whirlpool	<2–13	3.0	Kush and Hoadley (1980)
Whirlpool	<2–540	0.0–3.0	Kush and Hoadley (1980)
Whirlpool	2–49	0.0–0.2	Kush and Hoadley (1980)
Whirlpool	8–540	3.0	Kush and Hoadley (1980)
Whirlpool	81–920	No residual free chlorine detected	Kush and Hoadley (1980)
Whirlpools	90 to >180	In spite of routine disinfection	Hollyoak et al. (1995)
Whirlpools (9 of 9)	<1 to 10^8	In spite of routine disinfection	Price and Ahearn (1988)

Table 8 Diseases associated with *P. aeruginosa*'s presence in recreational or therapy water

Type of water	*P. aeruginosa* concentration	Type of disease	Number of affected persons	Reference
Swimming pool	Positive for bag of vacuum used to clean pool	Otitis externa	19	Reid and Porter (1981)
Swimming pool	Isolated from carpeting around pool	Folliculitis	14	Hopkins et al. (1981)
Swimming pool	Positive/ no quantitation	Skin rash	117	Thomas et al. (1985)
Swimming pool	Positive/no quantitation	Folliculitis	21	Ratnam et al. (1986)
Swimming pool	Positive/no quantitation	Folliculitis	37	Ratnam et al. (1986)
Swimming pool	Positive/no quantitation	Folliculitis	117	Ratnam et al. (1986)
Swimming pool	Positive/no quantitation	Earache	93	Thomas et al. (1985)
Two Whirlpools	Positive/no quantitation	Skin rash	27	MMWR (1979)
Whirlpool	9,000 CFU/100 mL	Urinary tract infection	2	Salmen et al. (1983)
Whirlpool	Positive/no quantitation	Urinary tract infection	1	Salmen et al. (1983)
Whirlpool	>180 CFU/100 mL	Wound infection	4	Hollyoak (1994)
Whirlpool	2.3×10^5 mL^{-1}	Pneumonia	1	Rose et al. (1983)
Whirlpool	Positive/no quantitation	Folliculitis	16	Hudson et al. (1985)
Whirlpool	No culture done	Folliculitis and corneal ulcer	12	Insler and Gore (1986)
Whirlpool	Positive/no quantitation	Folliculitis	42	Ratnam et al. (1986)
Whirlpool		Folliculitis	32	Washburn et al. (1976)
Whirlpool	Positive/no quantitation	Folliculitis	27	Ratnam et al. (1986)
Whirlpool	Positive/no quantitation	Folliculitis	20	Sausker (1978)
Whirlpool	Positive/no quantitation	Folliculitis	4	Ratnam et al. (1986)
Whirlpool	5.4×10^2 mL	Folliculitis	75	Ratnam et al. (1986)
Whirlpool	High concentrations	Folliculitis	4	Ratnam et al. (1986)
Whirlpool	Positive/no quantitation	Folliculitis	8	Ratnam et al. (1986)
Whirlpool	0.9×10^1 to 3.4×10^3 mL	Folliculitis	26	Ratnam et al. (1986)

(continued)

Table 8 (continued)

Type of water	P. aeruginosa concentration	Type of disease	Number of affected persons	Reference
Whirlpool	900/100 mL	Folliculitis	26	Ratnam et al. (1986)
Spa pool	8×10^6 to 1×10^7 mL^{-1}	Folliculitis	7	Ratnam et al. (1986)
Hot tub	High count	Folliculitis	6	Ratnam et al. (1986)
Hot tub	Positive/no quantitation	Folliculitis	6	Ratnam et al. (1986)
Hot tub	Positive/no quantitation	Folliculitis	5	Ratnam et al. (1986)
Hot tub	Positive/ no quantitation	Folliculitis	4	Ratnam et al. (1986)
Physiotherapy pool	Positive/no quantitation	Folliculitis	15	Ratnam et al. (1986)
Hydrotherapy pool	20 mL^{-1}	Chest infection of CF patients	2	Friend and Newsom (1986)
Water slide	Positive/no quantitation	Folliculitis	265	MMWR (1983)
Lake water	4 (1–63)/L	Ear infection	32	van Asperen et al. (1995)

Table 9 Presence of *P. aeruginosa* in disease-free recreational and surface waters

Type of water	Conc. of P. aeruginosa	Comments	Reference
Lake water	0– >1,000/100 mL	Recreational water	Seyfried and Cook (1984)
Lake water	0–100/100 mL	Recreational water	Seyfried and Cook (1984)
Lake water	0–10/100 mL	Recreational water	Seyfried and Cook (1984)
Lake water	0–10/100 mL	Recreational water	Seyfried and Cook (1984)
Lake water	31–524/100 mL	Recreational water	Falcao et al. (1993)
River water	33/100 mL		Seyfried and Cook (1984)
River water	268–536 CFU/L		Payment et al. (1988)
River water	34.5%		Pellett et al. (1983)
Creek water	1.7×10^6/100 mL		Highsmith and Abshire (1975)

Counts of *P. aeruginosa* taken from lake water, in one study, correlated with the numbers of bathers. At midnight when no bathers were present, no *P. aeruginosa* were detected; at 11 AM, when the number of bathers peaked, 2,400 *P. aeruginosa*/100 mL was detected. Similar serotypes have been isolated from lake water and ears of patients afflicted with otitis externa (Seyfried and Cook 1984). A summary of the occurrence of *P. aeruginosa* in recreational or natural waters not linked to disease is shown in Table 9.

6.4 Biofilms

Despite the view that bacteria are primitive organisms struggling for their individual survival, it is becoming clear that bacteria in nature seldom behave as isolated organisms (Kolter and Losick 1998). Most bacteria have evolved elaborate mechanisms for adhering to solid surfaces and for establishing biofilms. Bacterial biofilms are ubiquitous and serve many important roles in different environments. Biofilms have a specialized architecture that ensures the well-being of the individual cells that compose it. The sloughing off of individual cells from the biofilm completes the developmental cycle.

Microscopic observations of living bacterial biofilms attached to a glass surface have shown that biofilm populations have a complicated structural architecture. Biofilms of mixed bacterial communities, and of individual species such as *P. aeruginosa* that develop on solid surfaces exposed to a continuous flow of nutrients, form thick layers consisting of differentiated mushroom- and pillar-like structures separated by water-filled spaces (Davies et al. 1998). The structures consist primarily of an extracellular polysaccharide (EPS) matrix, or glycocalyx, in which the bacterial cells are embedded. This EPS matrix is considered to be important in cementing the bacterial cells together to form a biofilm.

Planktonic cells of *P. aeruginosa* and cells in biofilms produce similar amounts of EPS. However, the distribution of the glycocalyx is different, with biofilm cells cemented to one another by the EPS matrix, and planktonic cells having a compressed incomplete glycocalyx. Evidence suggests that without the lasI gene, the initial stages of biofilm formation proceed as normal, but differentiation from attached planktonic bacteria into biofilm bacteria does not proceed. This differentiation is possibly triggered when the cell mass produces a sufficient amount of the quorum-sensing signal, *N*-3-oxododecanoyl homoserine lactone (3OC12-HSL). Although the signal generated by *RhlI* does not appear to participate in biofilm differentiation, there may well be other, as yet unidentified, signals implicated in the process.

The finding that *P. aeruginosa* produces at least two extracellular signals involved in cell-to-cell communication and cell density-dependent expression of many secreted virulence factors, suggests that cell-to-cell signaling could be involved in the differentiation of *P. aeruginosa* biofilms.

Biofilms of wild-type *P. aeruginosa* cells are not sensitive to 0.2% sodium dodecyl sulfate (SDS), whereas biofilms composed of mutant cells deficient in the lasI product $3OC_{12}$-HSL are rapidly dispersed by SDS. The control of biofilm stability by certain gene products has important implications for the control of *P. aeruginosa* biofilms. Perhaps inhibition of the cell-to-cell signals could aid the treatment of biofilms (Davies et al. 1998). *P. aeruginosa* may be released, as part of a distribution system's biofilm, during certain hydraulic events. A city in Texas used a large volume of water to fight a major fire. The high water demand reduced line pressure. When pressure was restored, a biofilm sheared from a pipe section that entered the main waterline released 125–200 *P. aeruginosa* per milliliter over a 24-hr period, before again subsiding to nondetectable levels (Geldreich 1996).

6.5 *P. aeruginosa* from Other Sources

P. aeruginosa is broadly distributed in the environment. It has been isolated from plants and vegetables (Kominos et al. 1972). As part of the normal microbial flora, *P. aeruginosa* is sometimes present in humans; however, the prevalence of colonization in healthy humans is low (Pollack 1995). *P. aeruginosa* has been isolated from the stool of 1.2 to 2.3% of healthy persons (Botzenhart and Doring 1993), and colonization of the gut is thought to precede induction of disease (Pollack 1995).

Green et al. (1974) found *P. aeruginosa* in 24% of agricultural soil samples surveyed in California. Soils, in which tomatoes were grown, yielded the highest frequency of isolation. Of the positive samples, 10^0–10^3 CFU/g dry wt were found. *P. aeruginosa* was recovered from 0.19% of the samples taken from only 2 of the 43 surveyed fields. It was isolated from one tomato and one celery plant (counts were not determined).

Tomato salads prepared in a hospital kitchen contained *P. aeruginosa* 27% of the time at concentrations ranging from 10^0 to 10^3 mL^{-1} homogenate (Kominos et al. 1972). *P. aeruginosa* was found on 82% of the raw tomatoes at concentrations of 10^0–10^4 mL^{-1}, and on several other vegetables. The lowest rate of contamination was found for lettuce in which 10% of samples were positive; levels ranged from 10^0 to 10^2 CFU/g (Remington and Schimpff 1981).

Animals and insects also serve as reservoirs for *P. aeruginosa*. For example, one study showed that German cockroaches were potential vectors (Fotedar et al. 1993). At a high dose (10^7), *P. aeruginosa* was found to multiply and survive in the gut of the cockroaches and continue to be excreted for up to 114 d. However, Sundheim et al. (1998) examined 18 chicken carcasses from commercial poultry processing plants, and no *P. aeruginosa* was detected on any of them. An examination of 36 raw milk samples also failed to reveal the presence of *P. aeruginosa* (Shelley et al. 1987).

Although *P. aeruginosa* is sometimes found as normal flora in man, the prevalence of colonization in healthy persons outside of hospitals is low (2.6–25%; Hall et al. 1968; Jacobson 1985). Hospitalization leads to increased rates of carriage. Shooter (1971) showed that 24% of patients admitted to the hospital carried *P. aeruginosa* in the GI tract, and that at some time during their stay, 38% became carriers. Although the source of an infection is often difficult to trace, disease has been associated with contaminated hospital equipment. Contaminated foods and medicines have also been implicated as routes of transmission in hospitals (Shooter et al. 1969).

Contaminated milk may be a source of *P. aeruginosa* that is capable of causing outbreaks of diarrhea in a nursery for newborns (Hunter and Ensign 1947). In a German hospital, contaminated drinking water was the source of *P. aeruginosa* that caused funicular sepsis (infection of the umbilical cord resulting in bacteremia) in ten newborns (Weber et al. 1971).

6.6 Survival

P. aeruginosa can survive for several days in filtered freshwater that lacks suspended particles or other cellular microorganisms (de Vicente et al. 1988). The addition of 1% clarified sewage resulted in a 4-log reduction of *Pseudomonas* within a 28-d period. This decline probably resulted from competitors and predators in the sewage, because autoclaved sewage did not have the same effect. Similar tests with unfiltered freshwater also produced a 4-log decrease in the *P. aeruginosa* population over a 28-d period. Visible light resulted in a 3-log decrease over the same time period.

P. aeruginosa is able to grow at very low nutrient levels such as tap water (van der Kooij et al. 1982). When low numbers of *P. aeruginosa* was inoculated into bottled spring water, a 6-log increase in numbers was observed. The bacterium population was also found to increase in both tap and deionized water (Warburton et al. 1994). *P. aeruginosa* survived for very long periods when incubated in double-distilled water with *Salmonella*. The authors believed that this occurred because nutrients were released by *Salmonella*. Indeed, *P. aeruginosa* has been recovered in several mist therapy units, and has been shown to actually grow as a pure culture in distilled water (Favero et al. 1971; Botzenhart and Kufferath 1976). The ability of *P. aeruginosa* to survive in different kinds of water is summarized in Table 10.

7 Drinking Water Treatment for the Removal of *P. aeruginosa*

7.1 Iodine, Free Chlorine, and Chloramine

Drinking water is commonly treated by one of the following methods: chlorine, chloramines, ozone, iodine, and UV disinfection. Laboratory testing has not shown *P. aeruginosa* to be unusually resistant to chlorine, chloramine, ozone, or iodine. When iodine-resistant *P. aeruginosa*, isolated from a hospital povidone–iodine

Table 10 Survival of *P. aeruginosa* in various aqueous media

Type of water	Survival time (days)	Log change at end of test	Reference
Freshwater	>28	−3	de Vicente et al. (1988)
Groundwater	>12	−3.3	Chao et al. (1987)
River water	>8	−3	Chao et al. (1987)
Bottled spring water	>33	+7	Tamagnini and Gonzalez (1997)
Double distilled water with *Salmonella* present	>140	−0.1	Warburton et al. (1994)
Distilled water	>7	+4.4	Favero et al. (1971)
Mineral water	>20	−2	Moreira et al. (1994)
Sterile mineral water	>20	+0.1	Moreira et al. (1994)

solution, was tested against iodine, it was found to be more sensitive than several other tested bacteria: *P. fluorescens*, *P. cepacia*, *Bacillus* spp., and staphylococci (Pyle and McFeters 1989). Determining the sensitivity of *P. aeruginosa* to disinfectants is complicated by the fact that the sensitivity of the cells are affected by temperature, previous growth conditions, and the growth phase at the time of testing (Carson et al. 1972; Pyle and McFeters 1989). Cells are more rapidly killed when incubated at higher temperatures than at lower temperatures. They are also more sensitive to disinfection after being subcultured on laboratory media when being tested in a natural water environment. The die-off rate may be higher when log-phase populations are tested than when stationary phase cells are used.

In a study by Aspinall and Graham (1989), levels of free chlorine >1 ppm were sufficient to kill an initial count of 3.8×10^4 CFU/mL of *P. aeruginosa* after 10 sec of contact. However, these bacteria were isolated from a whirlpool and then grown in laboratory media before testing (Aspinall and Graham 1989). It required 1–3 hr to obtain a 99.9% kill in swimming pools with 0.5 mg/L chlorine at 25°C (Fitzgerald and DerVartanian 1969). These authors also demonstrated that natural waters required 0.5 mg of chlorine per liter to kill 99.9% of *P. aeruginosa* cells in 1 hr at 25°C.

Price and Ahearn (1988) demonstrated that *P. aeruginosa* cells regrew when the chlorine concentration decreased in whirlpools. *P. aeruginosa* has been recovered from whirlpool water containing 2 mg/L free chlorine. However, Highsmith et al. (1985) showed that *P. aeruginosa* test strains from whirlpool water were just as sensitive to low concentrations of chlorine (0.2–0.6 ppm) as control strains. This implied that chlorine resistance per se does not appear to be a characteristic of certain strains of *P. aeruginosa*. These bacteria can excrete glycocalyx material capable of shielding them from chlorine. There is no evidence that they are innately resistant to chlorine.

Vess et al. (1993) allowed *P. aeruginosa* to colonize PVC pipe material for 8 wk. The biofilm of pure *P. aeruginosa* was then exposed to 10–15 ppm free chlorine for 7 d during which time samples were taken at intervals to test for bacterial survival. After 7 d, the pipe was refilled with sterile distilled water and sampled at 7-d intervals for 6 wk for recovery and regrowth. Although no *P. aeruginosa* was detected at 7 d after the chlorine treatment, high numbers were recovered after 14 d. Scanning electron microscope pictures of the PVC wall surface showed cells embedded in heavy deposits of extracellular material, which probably accounted for the survival in the presence of chlorine.

The sensitivity of *P. aeruginosa* to chloramine was compared with *Klebsiella pneumoniae*, *Salmonella typhimurium*, and *E. coli* (Ward et al. 1984). *P. aeruginosa* was just as sensitive to chloramine, or even more so, than was the other bacteria. The time for 99.9% inactivation at 1 mg/L chloramine ranged from 4 to 9.5 min, with a more rapid activation at a pH of 6 than at pH 8. The time for 99.9% inactivation at 3 mg/L ranged from 1.5 to 4.5 min with the same influence of pH.

When *P. aeruginosa* was tested against *N*-chloramine (3-chloro-4,4-dimethyl-2-oxazolidnone), it was more sensitive than *Staphylococcus aureus* or *Shigella boydii* (Williams et al. 1985). In addition, at 2.5–5 ppm chlorine, no *P. aeruginosa* was recovered after 5–10 min at pH 7 and 22°C. At the same temperature and at a pH of 4.5, *P. aeruginosa* was not detectable after exposure to 2.5 ppm chlorine for 5–10 min, or exposure to 5.0 ppm for 2–5 min.

7.2 Other Disinfectants

P. aeruginosa has been shown to be no more resistant to ozonation than *Salmonella* or *Yersinia*. *P. aeruginosa* underwent a 6-log reduction in deionized water within 1 min, using 0.64–0.188 ppm ozone (Restaino et al. 1995). *P. aeruginosa* also appeared not to be more resistant to UV light than other bacteria, according to Box et al. (1982). *Bacillus subtilis* was most resistant to UV light at wavelengths from 250 to 270 nm, followed in order of decreasing resistance by *E. coli*, *P. aeruginosa*, and *S. aureus*. Fernandez and Pizarro (1996) also reported that *P. aeruginosa* was not resistant to UV light; in fact, it was more sensitive than *E. coli* in their studies. However, a comparison of the sensitivity of *P. aeruginosa* to several other bacteria was described by Wolfe (1990), who found that *P. aeruginosa* was more resistant than other bacteria tested (Table 11).

In recent years, copper and silver ions have been used to control *Legionella* and other water-based pathogens in water (Silvestry-Rodriquez et al. 2007a). *P. aeruginosa* appears to be susceptible to silver ions, but mutated forms of the organism may be more silver tolerant.

Silver requires longer contact to disinfect than does oxidizing disinfectants, but may augment the action of other disinfectants, where preventing colonization is important (e.g., faucets). Silver ions, at a concentration of 80 µg/L, have been reported to produce a 99.999% reduction in *P. aeruginosa* after 6 hr of contact (Huang et al. 2008). An advantage of silver is that it is effective in the presence of organic loads which render chlorine ineffective (Silvestry-Rodriquez et al. 2007b).

Table 11 Approximate dosages of UV light needed to inactivate 90% of selected bacteria

Bacterium	Dosage (µW-s)/cm^2
E. coli	3,000
S. typhi	2,500
P. aeruginosa	5,500
S. enteritidis	4,000
S. dysenteriae	2,200
S. paradysenteriae	1,700
S. flexneri	1,700
S. sonnei	3,000
S. aureus	4,500
L. pneumophila	380
Vibrio cholerae	3,400

8 Epidemiological Evidence for Transmission by Water

8.1 Hot Tubs, Spas, and Whirlpools

P. aeruginosa has caused more outbreaks of illness from contact with spas and hot tubs than any other organism (Gerba and Gerba 1998; Craun et al. 2005). Evidence points to improper maintenance as the major cause of such outbreaks, because where complete data on pool operation were available, illness was associated with inadequate pool disinfection and care. Common errors included failure to maintain adequate germicide concentrations, proper pH, and temperature and failure to change water or clean/replace filters.

Spas and hot tubs are pools designed for recreational and therapeutic use and for physiological and psychological relaxation. These pools are not drained, cleaned, and refilled after each use and may include devices such as hydrojet circulation, hot water, cold water, mineral baths, air induction systems, or some combination of these. Spas and hot tubs are shallow and not meant for swimming or diving. However, these facilities, like swimming pools, are closed cycle water systems and may be fitted to allow for complete water circulation, filtration, heating, and in some cases, disinfection and overflow protection (Gerba and Gerba 1998).

P. aeruginosa can quickly establish itself in a whirlpool spa if disinfection falls below recommended levels, even for a period as brief as 24 hr. For example, hoses used to fill whirlpools can be reservoirs for *P. aeruginosa*. Flushing of a used hose with tap water for 5–10 min reduced the densities of *P. aeruginosa* found in freshly filled whirlpools. The ability of *P. aeruginosa* to adhere tenaciously to PVC piping and filters makes complete removal of *P. aeruginosa* (biofilms) from a whirlpool system extremely difficult.

Factors contributing to outbreaks of skin infections from whirlpool contact include high temperature of the water conducive to organism growth, water turbulence and aeration, and heavy bather load per volume of water (Brett and du Vivier 1985; Ratnam et al. 1986). The longer a bather spends in the whirlpool, the greater the chances of developing *P. aeruginosa*- associated dermatitis (Hudson et al. 1985). Outbreaks may occur despite the presence of 0.5 ppm bromine (Ratnam et al. 1986).

Of the 17 serotypes of *P. aeruginosa* only 4, 6, 7, 9, 10, and 11 have been associated with outbreaks. Serotype 11 has been associated with over half of the outbreaks (Highsmith et al. 1985). The greater prevalence of *P. aeruginosa* serotype 11, among isolates from patients and pools led to the following hypotheses. (1) These strains of serotype 11 are particularly well adapted for survival and growth at high water temperatures, even in the presence of chlorine and (2) these strains possess a sufficient enzymatic complement of proteases and lechithinases so that, in the presence of heat and hydration, they can provoke an inflammatory reaction or infect the skin.

Table 12 Spa and hot tub disease outbreaks associated with *P. aeruginosa*

Type of illness	Number of outbreaks	Number of persons affected
Dermatitis (mild skin rash)	7	189
Folliculitis (severe skin rash)	22	492
Otitis externa (painful ear infection; swimmer's ear)	16	29
Keratitis (corneal ulcer)	1	2
Urinary tract infection (painful urination)	3	2
Total	49 (outbreaks)	714 (cases)

Gerba and Gerba (1998)

In a review of outbreaks, Gerba and Gerba (1998) cited 49 outbreaks involving 714 individuals (Table 12). The majority of outbreaks comprised skin rashes and infection of hair follicles (folliculitis), followed by ear infections (otitis externa). Eye and urinary tract infections have been less commonly reported (Insler and Gore 1986). The majority of outbreaks are recognized because the bathers are related (i.e., same family, sports club, health club, etc.) and communicate with each other when a common exposure results in illness (Jacobson 1985).

The first outbreak of whirlpool-associated *P. aeruginosa* dermatitis occurred in 1972 (McClausland and Cox 1975). This outbreak, like others reported later, occurred among whirlpool users exposed to *P. aeruginosa* serotype O:11. Generally, the contaminated whirlpools were adequately chlorinated (Khabbaz et al. 1983). This observation suggests that *P. aeruginosa* serotype O:11 may be either more invasive than other serotypes or better adapted to survive in chlorinated water. Another factor that may contribute to the pathogenesis of *Pseudomonas* spp. is the dilation of skin pores from the elevated temperature of whirlpool water (Khabbaz et al. 1983).

In an outbreak of dermatitis caused by *P. aeruginosa* serotype O:4, Hudson et al. (1985) first documented a relationship between duration of whirlpool use and development of *P. aeruginosa* dermatitis. The outbreak occurred among 16 guests at a hotel whirlpool spa. The attack rate among guests with 30 min or more of whirlpool spa use was 50% (13/26), in contrast to 13% (3/23) among guests with briefer exposures. The increased risk was observed for both one-time bathers and those who used the whirlpool more than once on the weekend. Logistic regression analysis also identified duration of whirlpool spa use as the variable which was most important in predicting illness among pool users. A high attack rate among females was believed to be associated with tighter fitting swimsuits that retained water near the skin after the swimmer left the pool.

Rose et al. (1983) described the case of a 47-yr-old man who contracted pneumonia after spending 90 min in his home whirlpool. The patient had nothing remarkable in his medical history except for a history of smoking. High numbers of *P. aeruginosa* were found in the whirlpool itself, and high numbers were also found in aerosols from the pool. When the pool was in operation, blood agar plates were opened, inverted, and the exposed agar surface was held 15 cm above, and facing the water surface for 30, 60, 90 and 180 s. The colony counts of *P. aeruginosa* on the plates were 22, 54, too numerous to count, and 90, respectively. It is obvious that aerosols from whirlpools may contain large numbers of *P. aeruginosa*.

8.2 Swimming Pools

Skin and eye infections are commonly associated with swimming pool use. *P. aeruginosa* is the second most common cause of recreational disease outbreaks (from swimming pools and hot tubs; Craun et al. 2005). Ear infection, or otitis externa, caused by *P. aeruginosa* are probably more common than reported. In one outbreak, 18 of 25 members of a competitive swimming team, who trained daily, developed painful discharges of the ear (Reid and Porter 1981). This group trained twice daily (early morning and late afternoon). Investigation revealed that chlorination was often inadequate during these times of day. In contrast, a group of competitive swimmers that trained in the early afternoon, when chlorine levels were adequate, were free of illness. In another instance, 265 persons developed *P. aeruginosa* folliculitis or ear infections associated with the use of a water slide over a 3-wk period (MMWR 1983). In this case, an inflatable plastic bubble covered the entire pool and deck areas. Within this "bubble" water and ambient air temperatures were 35°C and the relative humidity stood at 95%. Despite adequate disinfection, the infection rate was not resolved until the plastic bubble and surrounding indoor–outdoor carpeting were removed. In poorly maintained pools in Tehran, Iran, Hajjartabar (2004) found that *P. aeruginosa* could be isolated from the ear swabs of almost 80% of persons who reported ear problems, within 2 wk after swimming at the pool.

The use of water slides and play equipment has become more common in recent years, and such equipment may be prone to colonization by *P. aeruginosa*. In Wales, an outbreak of *Pseudomonas* folliculitis occurred after an outdoor gaming event took place; the event involved a water slide in which 151/593 (26%) of participating children were affected (Evans et al. 2003). The symptom was a rash which extended predominately over the lower truck and buttocks of those afflicted. The organism was isolated from the water and from a fire truck, which supplied the water.

8.3 Tap Water

Tap water is a significant source of nosocomial infections when used to prepare solutions, for hand washing or bathing, in bubble baths, in dialysis equipment, in waterbaths for heating solutions, and in other hospital equipment (Bert et al. 1998; Buttery et al. 1998; Doring et al. 1993; Ferroni et al. 1998; Fierer et al. 1967; Muyldermans et al. 1998; Vanholder et al. 1990). Although contaminated tap water was linked to these nosocomial outbreaks, contamination probably occurred from colonization of the tap hardware (Ferroni et al. 1998; Grundmann et al. 1993), aerosols generated by tap usage (Doring et al. 1993), sinks, or containers (Buttery et al. 1998; Fierer et al. 1967). Various studies have shown that between 9.7 and 67.1% of randomly taken tap water samples, from different intensive care units, were positive for *P. aeruginosa* (Trautmann et al. 2005).

Anaissie et al. (2002) estimated that 30% of all *P. aeruginosa* pneumonia infections in hospitals resulted from contact with tap water; these contacts resulted in 1,400 deaths per year in the United States. Exposure probably occurs from colonization of tap hardware and associated plumbing fixtures by the bacterium. Patients are exposed while showering, bathing, and drinking water or ice, and through contact with medical equipment (e.g., tube feed bags, endoscopes, and respiratory equipment) rinsed with contaminated tap water. Studies usually have failed to identify the organism in the water distribution system, so the original source of the organism is not clear. Some believe that the source exists in end pipes and biofilms within the distribution system, and eventually results in colonization of fixtures. As mentioned earlier, *P. aeruginosa*, found in a contaminated water supply, has been linked to one waterborne outbreak of funicular sepsis (infection of the umbilical cord resulting in bacteremia). The infection occurred in ten newborns held in a hospital nursery in Germany (Weber et al. 1971). The groundwater supply was found to be contaminated by seepage of sewage and infiltration of contaminated surface water.

Recent hospital intervention studies have further demonstrated the significance of tap water as a source of nosocomial infections. Use of POU filters on taps in intensive care units and other locations, where sensitive populations are cared for, has been shown to significantly reduce *P. aeruginosa* infections and colonization of patients by the organism (Vianelli et al. 2006; Trautmann et al. 2008). Trautmann et al. (2008) reported that the mean monthly rate of *P. aeruginosa* infections/colonizations in an intensive care unit decreased from an average of 3.9 to 0.8 after the installation of 0.2-μm disposal water filters on the taps.

A study of several families, who use contaminated well water, has suggested an association between showers/baths and *P. aeruginosa* infection rates (Zichichi et al. 2000). The organism was in the well water of four families, whereas it was detected in the shower or bath rug of the other families. The authors suggest that *P. aeruginosa* folliculitis infection may be more common than actually diagnosed, because it may be confused with insect bites, scabies, and other events of infectious folliculitis. Another outbreak attributed to *P. aeruginosa* caused gastroenteritis in Mexican children, and was associated with contaminated groundwater (de Victoria and Galvan 2001).

An outbreak of *Pseudomonas* folliculitis (skin rashes) was reported among workers in a cardboard manufacturing plant shortly after the facility switched to a closed-loop water recycling system (Hewitt et al. 2006). This switch resulted in overgrowth of *P. aeruginosa* in the water system. In another study, Jordanians who drank unchlorinated water were more likely to be colonized by *P. aeruginosa* than those who drank from chlorinated supplies (15 vs. 44%; Shehabi et al. 2005). This study also demonstrated similar antimicrobial resistance patterns, biotypes, and serotypes among the *P. aeruginosa* from patients' stools compared to isolates from the drinking water sources.

9 Water Quality Standards and Guidelines for *P. aeruginosa*

Various individuals and government entities have recommended standards to prevent infection by *P. aeruginosa*. Cliver et al. (1987) suggested that *P. aeruginosa* should be absent in 100-mL volumes of finished drinking water since it is a potential pathogen. In 1995, the European Union published a draft Drinking Water Directive which included a maximum permissible standard for *P. aeruginosa* of zero per 250 mL of bottled water (EU 1995).

Hoadley (1977) stressed the need for regulation of whirlpool baths and maintenance of water quality. His studies suggested that existing regulations promulgated for swimming pools (relating to coliform counts) were of little value for protecting bathers at whirlpool baths. Broadbent (1996) suggested a limit of no detectable *P. aeruginosa* in a 100 mL sample in spas, hot tubs, and whirlpools. This author further recommended that public spas be tested weekly during their first month of operation and then monthly thereafter, if test results indicate that the water is safe. Martins et al. (1995) conducted an extensive study of water quality in swimming pools, and evaluated various microbial indicators, including *Canadida albicans*, *S. aureus*, total coliforms, fecal coliforms, fecal streptococcus, heterotrophic plate count (HPC) bacteria, and *P. aeruginosa*. *C. albicans* and *P. aeruginosa* were the only organisms that did not correlate with the other indicators, and it was suggested that they were not useful indicators of human contamination. *P. aeruginosa* was the organism that was least detected. In a case-controlled epidemiological study, van Asperen (1995) identified a significant risk of otitis externa (ear infections) from exposure to *P. aeruginosa* in freshwater lakes that met current Dutch and European Commission coliform standards.

10 Risk Assessment

Considering what is currently known, it is probable that most of the heterotrophic bacteria in drinking water are not human pathogens. However, some of the genera, such as *Pseudomonas*, *Legionella*, and *Mycobacterium*, include species that are opportunistic pathogens. The same is true of the following genera that may be included as part of the standard HPC bacteria: *Acinetobacter*, *Xanthomonas*, *Moraxella*, and *Aeromonas* (Rusin et al. 1997).

An opportunistic pathogen is defined as one that usually causes disease only when the host immune system is weakened, or when the pathogen overcomes other body barriers (i.e., the skin). Study results suggest that *Pseudomonas* is of concern when it penetrates the skin, whether as a result of prolonged contact with water or entry through burn wounds. With the increase in the immunocompromised population from AIDS, organ transplantation, and chemotherapy, there has been an increased interest in and concern for opportunistic pathogens. In this context, the purpose of this section is to address the risk assessment of human ingestion of *P. aeruginosa* in drinking water.

Where sufficient data existed, the following four-tiered approach was used to perform this risk assessment (1) *hazard identification* – identification of the causative microbe and the spectrum of human diseases associated with this microorganism, (2) *exposure assessment* – determination of the opportunistic pathogen concentration in drinking water (exposure), (3) *dose–response assessment* – connects the relationship between administered dose and the probability of infection or disease in the exposed population, and (4) *risk characterization* – estimation of the potential impact of the opportunistic pathogen based on the severity of its effects and degree of exposure. Daily risks were estimated based on the consumption of 2 L/d of drinking water.

The development of a quantitative dose–response relationship is a primary step in performing a risk analysis. In assessing microbial risk, the dose–response relationship enables estimates to be made on the likelihood of an infection occurring, although some dose–response studies focus on colonization as an outcome rather than infection.

10.1 Occurrence of *P. aeruginosa* in Drinking Water

As reported previously, between <2.0 and 24.1% of tap water samples studied contained *P. aeruginosa*; levels detected ranged from 0 to 2,300 CFU/mL. This bacterium has been found in water vending machine samples with a frequency of 16.1%, and at concentrations of 0.002–0.16 CFU/mL; in bottled water the frequency was <0.4–18.8%, and concentrations varied from 0 to 10 CFU/mL.

10.2 Infective Dose

The development of a quantitative dose–response relationship is a primary step in performing a risk assessment (Rusin et al. 1997). The infective dose for *P. aeruginosa* is quite high, according to laboratory studies by George et al. (1989a). In this study, mice were dosed by gavage with 10^9 CFU of *P. aeruginosa*. None of the mice showed any signs of illness, and in 75% of the cases, no colonization of the gut took place, unless the mice were repeatedly inoculated by eating their own feces. When mice were colonized, *P. aeruginosa* was subsequently isolated from their feces for up to 14 d. The *P. aeruginosa* did not cause detectable disease as measured by appetite, activity, overall appearance, and weight.

A high infective dose for mice was also found for aerosols (George et al. 1991). After an intranasal inoculation of 1.6×10^3 CFU/animal, *P. aeruginosa* was readily cleared from the body. No mortality and no observable signs of morbidity were evident. After inoculation with 1.6×10^7 CFU, *P. aeruginosa* was detectable in the GI tract within 3 hr. The bacteria persisted in the small and large intestine and in the cecum for 14 d; lethargy and weight loss were observed in these mice. By the intranasal route, the LD_{50} in mice for *P. aeruginosa* was 2.7×10^7 CFU. Mice inoculated with 2.2×10^9 CFU died within 24–36 hr.

Hentges et al. (1985) showed that antibiotics decreased the resistance of mice to intestinal colonization, when they were inoculated orally with 10^8 CFU *P. aeruginosa*. Among mice that did not receive antibiotics, 20% still passed *P. aeruginosa* in the feces on day 14, as compared to mice treated with ampicillin (90%), clindamycin (70%), and metronidazole (50%). George et al. (1989b) also showed that ampicillin increased the recovery of *P. aeruginosa* from the GI tract of mice, compared to untreated animals.

Buck and Cooke (1969) found that colonization of healthy human volunteers by *P. aeruginosa* required an oral dose $\geq 1.5 \times 10^6$ CFU. With oral doses of 1.5×10^6 to 2.0×10^8, excretion in the feces was subsequently detected for up to 14 d, if the volunteer was also taking ampicillin. Excretion was limited to 6 d by volunteers not taking antibiotics. None of the volunteers experienced any disease symptoms from the *P. aeruginosa*. Table 13 summarizes the studies of Buck and Cooke (1969) and Hentges et al. (1985).

10.3 Appraisal of Risk

By using the risk assessment approach of hazard identification, exposure assessment through ingestion of drinking water, and dose–response modeling, a risk characterization was developed to estimate the probability of infection for individuals consuming potable water containing *P. aeruginosa*. The exponential model provided the best fit with no improvement in fit if evaluated with the beta-Poisson model (a two-parameter model). The exponential model is as follows:

$$P_i = 1 - e^{(-[1/k] \times [N])},$$

where P_i is the probability of infection, $1/k$ is the fraction of ingested microorganisms that survive to initiate infection, and N is the number of microorganisms ingested. It was necessary to employ dose–response data from experiments conducted at high doses, fit mathematical models to these data sets, and extrapolate to doses representative of actual environmental exposure. Daily risks were calculated based on daily individual water consumption of 2 L/d. Risks were calculated for low

Table 13 Estimates of infectious doses of *P. aeruginosa*

Bacterium	Animal	Route of exposure	Median dose
Pseudomonas aeruginosa	Humans[a]	Oral	10^{10}
	Humans – ampicillin[a]	Oral	10^7
	Mice[b]	Oral	10^8
	Mice – antibiotics[b]	Oral	$10^4 – 10^8$

[a]Buck and Cooke (1969)
[b]Hentges et al. (1985)

Table 14 Calculated daily risks of colonization by *Pseudomonas aeruginosa* ingested in drinking water

Bacterium	Colony-forming units/2L/ person daily exposure	Daily risk of colonization for single exposure
Pseudomonas aeruginosa	20	1.7×10^{-8}
	4,600,000	3.8×10^{-3}
Pseudomonas aeruginosa – ampicillin	20	4.1×10^{-7}
	4,600,000	9.0×10^{-2}

water concentrations (10 CFU/L) and high water concentrations (2,300,000 CFU/L) of *P. aeruginosa* (Table 14).

10.4 Risks Associated with *P. aeruginosa* in Drinking Water

Risks were calculated based on a single exposure. We emphasize that the models predict colonization rather than disease. It is unclear whether colonization of the intestinal tract is a prerequisite for pneumonia, meningitis, or septicemia.

Conclusions based on this risk assessment are:

1. The data for ampicillin-pretreated subjects showed a possibility of a shift to greater sensitivity with pretreatment (Table 14). However, the numbers of subjects and doses were too low to give significant statistical power. Further research should clarify the impact of antibiotic treatment on dose–response relationships.
2. In the absence of disease, the risk for colonization when exposed to high levels of *P. aeruginosa* was predicted to be 9×10^{-2} [9/100], with antibiotic pretreatment (Table 14).
3. Sensitivity analysis of daily risks showed that the dominant source of variability/uncertainty was in the density distribution rather than in the dose–response or water consumption distributions. The sensitivity analysis of daily risks showed the following percentage contribution to variance (a) density of bacterium per milliliter of drinking water, 95.5%; (b) dose–response, 3.6%; and (c) daily water consumption, 1.0%. In most studies, no quantitative information was available, and often the species or even the genus was not identified. Therefore, the greatest need is for occurrence data to be both quantitative and specific regarding genus and species. There is also a great need for data on infectious dose, particularly with different serotypes of *P. aeruginosa*, and identification of strains possessing various virulence factors. The greatest health risk from water exposure appears to be through the skin, in hot tubs, or in the lungs by aerosols.
4. Overall, the risk presented by *P. aeruginosa* is seen as being transient, with the probability of infection being low; the chance of a severe outcome is associated primarily with the hospitalized patient or those with a weakened immune system. At this time, the disease outcome in these vulnerable populations cannot be quantified.

11 Data Gaps

This review suggests that tap water and bathing waters contaminated with high numbers of *P. aeruginosa* can cause infections in both compromised and normal, healthy individuals. In healthy individuals, the risk is probably limited to those who bathe (skin infections) and wear contact lenses. What remains unclear is the origin of the organism in most of the outbreaks. The limited data suggest that tap water becomes contaminated by the growth of *P. aeruginosa* in the faucet or shower head, but it is not known if the organism originated from the person using the tap or shower, or from the water itself. This is also true for hot tubs, spas, and swimming pools.

Addressing the following data gaps would allow for more comprehensive and dynamic quantitative microbial risk assessments for *Pseudomonas* spp.:

1. Defining the origin of *Pseudomonas* spp. in various water sources using molecular fingerprinting techniques (Grundmann et al. 1993)
2. Obtaining quantitative and species-specific data on the occurrence of *Pseudomonas* in distribution systems and tap waters for informed exposure assessments
3. Obtaining occurrence data for *Pseudomonas* spp. in undisinfected groundwater and in baths/showers/taps, since at least one study suggests that *P. aeruginosa* skin infections are a common occurrence in households which use contaminated, undisinfected groundwater (Zichichi et al. 2000). This would allow for a comparison of exposures from disinfected water systems
4. Investigating whether gastrointestinal colonization in normal, healthy individuals precedes infection in other parts of the body
5. Comparing dose–response data from single and multiple oral doses in animals given antibiotics to allow for the assessment of the effect of antibiotics on infective dose. Colonization increases among patients who are hospitalized or are taking antibiotics, and greater colonization of the gut could allow for greater dissemination in hospitals
6. Obtaining dose–response data regarding skin infections so that bathing as an exposure could be addressed in a quantitative microbial risk assessment

12 Summary

P. aeruginosa is part of a large group of free-living bacteria that are ubiquitous in the environment. This organism is often found in natural waters such as lakes and rivers in concentrations of 10/100 mL to >1,000/100 mL. However, it is not often found in drinking water. Usually it is found in 2% of samples, or less, and at concentrations up to 2,300 mL^{-1} (Allen and Geldreich 1975) or more often at 3–4 CFU/mL. Its occurrence in drinking water is probably related more to its ability to colonize biofilms in plumbing fixtures (i.e., faucets, showerheads, etc.) than its presence in the distribution system or treated drinking water. *P. aeruginosa* can survive in deionized or distilled water (van der Kooij et al. 1982; Warburton et al. 1994).

Hence, it may be found in low nutrient or oligotrophic environments, as well as in high nutrient environments such as in sewage and in the human body.

P. aeruginosa can cause a wide range of infections, and is a leading cause of illness in immunocompromised individuals. In particular, it can be a serious pathogen in hospitals (Dembry et al. 1998). It can cause endocarditis, osteomyelitis, pneumonia, urinary tract infections, gastrointestinal infections, and meningitis, and is a leading cause of septicemia. *P. aeruginosa* is also a major cause of folliculitis and ear infections acquired by exposure to recreational waters containing the bacterium. In addition, it has been recognized as a serious cause of keratitis, especially in patients wearing contact lenses. *P. aeruginosa* is also a major pathogen in burn and cystic fibrosis (CF) patients and causes a high mortality rate in both populations (Molina et al. 1991; Pollack 1995).

P. aeruginosa is frequently found in whirlpools and hot tubs, sometimes in 94–100% of those tested at concentrations of <1 to 2,400 CFU/mL. The high concentrations found probably result from the relatively high temperatures of whirlpools, which favor the growth of *P. aeruginosa*, and the aeration which also enhances its growth. The organism is usually found in whirlpools when the chlorine concentrations are low, but it has been isolated even in the presence of 3.00 ppm residual free chlorine (Price and Ahearn 1988). Many outbreaks of folliculitis and ear infections have been reportedly associated with the use of whirlpools and hot tubs that contain *P. aeruginosa* (Ratnam et al. 1986). Outbreaks have also been reported from exposure to *P. aeruginosa* in swimming pools and water slides.

Although *P. aeruginosa* has a reputation for being resistant to disinfection, most studies show that it does not exhibit any marked resistance to the disinfectants used to treat drinking water such as chlorine, chloramines, ozone, or iodine. One author, however, did find it to be slightly more resistant to UV disinfection than most other bacteria (Wolfe 1990). Although much has been written about biofilms in the drinking water industry, very little has been reported regarding the role of *P. aeruginosa* in biofilms.

Tap water appears to be a significant route of transmission in hospitals, from colonization of plumbing fixtures. It is still not clear if the colonization results from the water in the distribution system, or personnel use within the hospital. Infections and colonization can be significantly reduced by placement of filters on the water taps.

The oral dose of *P. aeruginosa* required to establish colonization in a healthy subject is high (George et al. 1989a). During dose–response studies, even when subjects (mice or humans) were colonized via ingestion, there was no evidence of disease. *P. aeruginosa* administered by the aerosol route at levels of 10^7 cells did cause disease symptoms in mice, and was lethal in aerosolized doses of 10^9 cells. Aerosol dose–response studies have not been undertaken with human subjects.

Human health risks associated with exposure to *P. aeruginosa* via drinking water ingestion were estimated using a four-step risk assessment approach. The risk of colonization from ingesting *P. aeruginosa* in drinking water is low. The risk is slightly higher if the subject is taking an antibiotic resisted by *P. aeruginosa*. The fact that individuals on ampicillin are more susceptible to *Pseudomonas* gastrointestinal infection probably results from suppression of normal intestinal flora, which would allow *Pseudomonas* to colonize. The process of estimating risk was significantly constrained because of the absence of specific (quantitative) occurrence data for *Pseudomonas*. Sensitivity analysis shows that the greatest source of variability/uncertainty in the risk assessment is from the density distribution in the exposure rather than the dose–response or water consumption distributions. In summary, two routes appear to carry the greatest health risks from contacting water contaminated with *P. aeruginosa* (1) skin exposure in hot tubs and (2) lung exposure from inhaling aerosols.

References

Adlard PA, Kirov SM, Sanderson K, Cox GE (1998) *Pseudomonas aeruginosa* as a cause of infectious diarrhea. Epidemiol Infect 121:237–241.

Agger WA, Mardan A (1995) *Pseudomonas aeruginosa* infections of intact skin. Clin Infect Dis 20:302–308.

Aksamit TR (1992) *Pseudomonas* pneumonia and bacteremia in the immunocompromised patient. In: Fick R (ed) *Pseudomonas aeruginosa* – The opportunist pathogenesis and disease. CRC, Boston, MA, pp. 177–180.

Allen MJ, Geldreich EE (1975) Bacteriological criteria for groundwater quality. Ground Water 13:45–52.

Alonso JL, Garay E, Hernandez E (1989) Membrane filter procedure for enumeration of *Pseudomonas aeruginosa* in water. Water Res 23:1499–1502.

Amromin GD, Solomon RD (1962) Necrotizing enteropathy a complication of treated leukemia or lymphoma patients. JAMA 182:23–29.

Anaissie, EJ, Penzak SR, Dignani MC (2002) The hospital water supply as a source of nosocomial infections. Arch Intern Med 162:1483–1492.

Angel C, Patrick CC, Lobe T, Rao B, Pui CH (1991) Management of anorectal/perineal infections caused by *Pseudomonas aeruginosa* in children with malignant diseases. J Pediatr Surg 26:487–492.

Anon (1994) *Pseudomonas*. In: Holt JG, Krieg NR, Sneath PHA, Staley JT, Williams ST (eds) Bergey's manual of determinative bacteriology, 9th Ed. Williams and Wilkins, Baltimore, MD, pp. 93–94.

Artenstein A, Cross A (1993) Local and disseminate diseases caused by *Pseudomonas aeruginosa*. In: Campa M, Bendinelli M, Friedman H (eds) *Pseudomonas aeruginosa* as an opportunistic pathogen. Plenum, New York, NY, pp. 223–239.

Asboe D, Gant V, Aucken HM (1998) Persistence of *Pseudomonas aeruginosa* strains in respiratory infection in AIDS patients. AIDS 12:1771–1775.

Aspinall ST, Graham R (1989) Two sources of contamination of a hydrotherapy pool by environmental organisms. J Hosp Infect 14:285–292.

Baumgartner A, Grand M (2006) Bacteriological quality of drinking water from dispensers (coolers) and possible control measures. J Food Protect 69:3043–3046.

Berk R, Brown D, Coutinho I, Meyers D (1987) In vivo studies with two phospholipase C fractions from *Pseudomonas aeruginosa*. Infect Immun 55:1728–1730.

Bert F, Maubec E, Bruneau B, Berry P, Lambert-Zeeovsky N (1998) Multi-resistant *Pseudomonas aeruginosa* outbreak associated with contaminated tap water in a neurosurgery intensive care unit. J Hosp Infect 39:53–62.

Bishop JF, Schimpff SC, Diggs CH, Wiernik PH (1981) Infections during intensive chemotherapy for non-Hodgkin's lymphoma. Ann Intern Med 95:549–555.

Bobo RA, Newton EJ, Jones LF (1973) Nursery outbreak of *Pseudomonas aeruginosa*: Epidemiological conclusions from five different typing methods. Appl Microbiol 25:414–420.

Bodey G, Bolivar R, Fainstein V, Jadeja L (1983) Infections caused by *Pseudomonas aeruginosa*. Rev Infect Dis 5:279–313.

Bodey GP, Jadeja L, Elting L (1985) *Pseudomonas* bacteremia: Retrospective analysis of 410 episodes. Arch Intern Med 145:1621–1629.

Botzenhart K, Doring G (1993) Ecology and epidemiology of *Pseudomonas aeruginsoa*. In: Campa M, Bendinelli M, Friedman H (eds) *Pseudomonas aeruginosa* as an opportunistic pathogen. Plenum, New York, NY, pp. 3–8.

Botzenhart K, Kufferath R (1976) On the growth of various Enterobacteriaceae, *Pseudomonas aeruginosa*, and *Alkaligenes* spp. in distilled water, de-ionized water, tap water and mineral salt solution. Zbl Bakt Hyg I Abt Orig B 163:470–485.

Box JA, Sugden JK, Younis NMT (1982) The use of ultraviolet light to sterilize water. Pharm Acta Helv 57:330–333.

Brett J, du Vivier A (1985) *Pseudomonas aeruginosa* and whirlpools. Br Med J 290:1024–1025.

Broadbent C (1996) Guidence on water quality for heated spas. South Australian Health Commission, National Environmental Health Forum, Rundle Mall, South Australia.

Bryan CS, Reynolds KL (1984) Bacteremic nosocomial pneumonia. Analysis of 172 episodes from a single metropolitan area. Am Rev Respir Dis 129:668–671.

Buck AC, Cooke EM (1969) The fate of ingested *Pseudomonas aeruginosa* in normal persons. J Med Microbiol 2:521–525.

Buttery JP, Alabaster SJ, Heine RG, Scott SM, Cruchfield, RA, Bigham A, Tabrizi SN, Garland SM (1998) Multiresistant *Pseudomonas aeruginosa* in a pediatric oncology ward related to bath toys. Pediatr Infect Dis 17:509–513.

Calderon R, Mood EW (1982) An epidemiological assessment of water quality and swimmer's ear. Arch Environ Health 37:300–305.

Carson LA, Favero MS, Bond WW, Petersen NJ (1972) Factors affecting comparative resistance of naturally occurring and subcultured *Pseudomonas aeruginosa* to disinfectants. Appl Microbiol 23:863–869.

Celis R, Torres A, Gatell J, Alnela M (1988) Nosocomial pneumonia: A multivariate analysis of risk and prognosis. Chest 93:318–322.

Chaidez C, Rusia P, Naranjo J, Gerba CP (1999) Microbiological quality of water vending machines. Int J Environ Health Res 9:197–206.

Chao WL, Ding RJ, Chen RS (1987) Survival of pathogenic bacteria in environmental microcosms. Chin J Microbial Immunol 20:339–348.

Chernick N, Armstrong D, Posner JB (1973) Central nervous system infections in patients with cancer. Medicine 52:563–581.

Clark JA, Burger CA, Sabatinos LE (1982) Characterization of indicator bacteria in municipal raw water, drinking water, and new main water samples. Can J Microbiol 28:1002–1013.

Cliver DO, Newman RA, Cortruvo JA (1987) Drinking water microbiology. J Environ Pathol Toxicol Oncol 7:1–365.

Craun GF, Calderon RL, Craun MF (2005) Outbreaks associated with recreational water in the United States. Int J Environ Health Res 15:243–262.

Davies DG, Parsek MR, Pearson JP (1998) The involvement of cell-to-cell-signals in the development of a bacterial biofilm. Science 280:295–298.

de Vicente A, Aviles M, Borrego JJ, Romero P (1988) Die-off and survival of *Pseudomonas aeruginosa* in freshwater. Zbl Bakt Hyg B 185:534–547.

de Victoria J, Galvan M (2001) *Pseudomonas aeruginosa* as an indicator of health risk in water for human consumption. Water Sci Technol 43:49–52.

Dembry LM, Zervos MJ, and Hierholzer WJ (1998) Nosocomial bacterial infections. In: Evans AS, Brachman PS (eds) Bacterial infections of humans: Epidemiology and control, 3rd Ed. Plenum Medical Book Company, New York, NY, pp. 509–511.

Dinapoli R, Thomas J (1971) Neurologic aspects of malignant external otitis: Report of three cases. Mayo Clin Proc 46:339–344.

Doring G, Bareth H, Gairing A (1989) Genotyping of *Pseudomonas aeruginosa* sputum and stool isolates from cystic fibrosis patients: Evidence for intestinal colonization and spreading into toilets. Epidemiol Infect 103:555–564.

Doring G, Ulrich M, Muller W (1991) Generation of *Pseudomonas aeruginosa* aerosols during hand washing from contaminated sink drains, transmission to hands of hospital personnel, and its prevention by use of a new heating device. Zbl Hyg 191:494–505.

Doring G, Horz M, Ortelt J, Grupp H, Wolz C (1993) Molecular epidemiology of *Pseudomonas aeruginosa* in an intensive care unit. Epidemiol Infect 110:427–436.

Duquino HH, Rosenberg FA (1986) Antibiotic-resistant *Pseudomonas* in bottled drinking water. Can J Microbiol 33:286–289.

Eng R, Bishburg E, Smith S (1986) Bacteremia and fungemia in patients with acquired immune deficiency syndrome. Am J Clin Pathol 86:105–107.

Ensign PR, Hunter CA (1946) An epidemic of diarrhea in the newborn nursery caused by a milk-borne epidemic in the community. J Pediatr 29:620–628.

EU (1995) Proposal for Council Directive concerning the quality of water intended for human consumption. Com (94) 612 Final 131:5–24.

Evans MR, Wilkinson EJ, Jones R, Mathias K, Lenartowicz P (2003) Presumed *Pseudomonas* folliculitis outbreak in children following an outdoor games event. Commun Dis Public Health 6:18–21.

Faden H (1998) Monthly prevalence of Group A, B, and G i types E and F, and *Pseudomonas aeruginosa* nasopharyngeal colonization in the first two years of life. Pediatr Infect Dis J 17:255–256.

Fagon JY, Cahstre J, Domart Y (1989) Nosocomial analysis of 52 episodes with use of a protected specimen brush and quantitative culture techniques. Am Rev Respir Dis 139:877–884.

Falcao DP, Leite CQF, Silva MJ (1993) Microbiological quality of recreational waters in Araraquara, SP, Brazil. Sci Total Environ 128:37–49.

Farmer III JJ (1995) Enterobacteriaceae: Introduction and identification. In: Murray PR (editor-in-chief) Manual of clinical microbiology, 6th Ed. ASM, Washington, DC, p. 443.

Farrell PM, Shen G, Splaingard M (1997) URL: http://www.pediatrics.org/cgi/content/full/100/5/e2.

Favero M, Carson L, Bond W, Petersen N (1971) *Pseudomonas aeruginosa*: Growth in distilled water from hospitals. Science 173:836–838.

Fernandez RO, Pizarro RA (1996) Lethal effect induced in *Pseudomonas aeruginosa* exposed to ultraviolet-A radiation. Photochem Photobiol 64:334–339.

Ferroni A, Nguyen L, Pron B, Quesne G, Brusset MC, Berche P (1998) Oubreak of nosocomial urinary tract infections due to *Pseudomonas aeruginosa* in a pediatric surgical unit associated with tap-water contamination. J Hosp Infect 39:301–307.

Fick RB (1992) *Pseudomonas aeruginosa* – the microbial hyena and its role in disease: An introduction. In: Fick R (ed) *Pseudomonas aeruginosa* – The opportunist pathogenesis and disease. CRC, Boston, MA, pp. 1–4.

Fierer J, Taylor PM, Gezon HM (1967) *Pseudomonas aeruginosa* epidemic traced to delivery-room resuscitators. N Engl J Med 276:991–996.

Fitzgerald G, DerVartanian M (1969) *Pseudomonas aeruginosa* for the evaluation of swimming pool chlorination and algicides. Appl Environ Microbiol 17:415–421.

Fletcher EL, Fleissig SMJ, Brennan NA (1993) Lipopolysaccharide in adherence of *Pseudomonas aeruginosa* to the cornea and contact lenses. Invest Ophthalmol Visual Sci 34:1930–1935.

Flores G, Stavola JJ, Noel GJ (1993) Bacteremia due to *Pseudomonas aeruginosa* in children with AIDS. Clin Infect Dis 16:706–708.

Fong I, Tomkins B (1985) Review of *Pseudomonas aeruginosa* meningitis with special treatment with ceftazidime. Rev Infect Dis 7:604–612.

Fotedar R, Banerjee U, Shriniwas S (1993) Vector potential of the German cockroach in dissemination of *Pseudomonas aeruginosa*. J Hosp Infect 23:55–59.

Friend PA, Newsom SWB (1986) Bacterial resistance to antiseptics and disinfectants. J Hosp Infect 7:213–225.

Gambassini L, Sacco C, Lanciotti E (1990) Microbial quality of the water in the distribution system of Florence. Aqua 39:258–264.

Garibaldi RA (1985) Epidemiology of community-acquired respiratory tract infections in adults. Am J Med 78(Suppl. 6B):32–37.

Geldreich EE (1996) Microbial quality of water supply in distribution systems. CRC, Boca Raton, FL, NY, p. 293.

George SE, Kohan MJ, Walsh DB, Claxton LD (1989a) Acute colonization study of polychlorinated biphenyl-degrading Pseudomonads in the mouse intestinal tract: Comparison of single and multiple exposures. Environ Toxicol Chem 8:123–131.

George SE, Walsh DB, Stead AG, Claxton LD (1989b) Effect of ampicillin-induced alterations in murine intestinal microbiota on the survival and competition of environmentally released pseudomonads. Fund Appl Toxicol 13:670–680.

George SE, Kohan MJ, Whitehouse DA (1991) Distribution, clearance, and mortality of environmental Pseudomonads in mice upon intranasal exposure. Appl Environ Microbiol 57:2420–2425.

Gerba CP, Gerba P (1998) Outbreaks caused by *Pseudomonas aeruginosa* associated with whirlpool spas, hot tubs, and swimming pools. Proceedings of the second annual chemistry symposium national spa and pool institute. National spa and pool institute, Alexandria, VA, pp. 8–18.

Gilligan P (1995) *Pseudomonas* and *Burkholderia*. In: Murray PR, Baron EJ, Pfaller MA, Tenover FC, Yolken, RH (eds) Manual of clinical microbiology, 6th Ed. ASM, Washington, DC, pp. 509–519.

Gilligan PH, Lum G, VanDamme PAR, Whittier S (2003) *Burkholderia, Stenotrophomonas, Ralstonia, Brevundimonas, Comamonas, Delftia, Pandoraea*, and *Acidivorax*. In: Murray PR, Baron EJ, Jorgesen JH (eds) Manual of clinical microbiology, 8th Ed. ASM, Washington, DC, pp. 729–748.

Goldman M, Rosenfel-Yehoshua N, Lerner-Geva L, Lazarovitch T, Schwartz D, Grisaru-Soen G (2008) Clinical features of community-acquired *Pseudomonas aeruginosa* urinary tract infections in children. Pediatr Nephrol 23:765–768.

Green SK, Schroth MN, Cho JJ, Kominos SD, Vitanza-Jack VB (1974) Agricultural plants and soil as a reservoir for *Pseudomonas aeruginosa*. Appl Microbiol 28:987–991.

Greenman RL, Schein RMH, Martin MA (1991) A controlled clinical trial of E5 murine monoclonal IgM antibody to endotoxin in the treatment of gram-negative sepsis. JAMA 266:1097–1102.

Gregory DW, Schaffner W (1987) *Pseudomonas* infections associated with hot tubs and other environments. Infect Dis Clinics North Am 1:635–648.

Grundmann H, Kropec A, Hartung D, Berner R, Daschner F (1993) *Pseudomonas aeruginosa* in a neonatal intensive care unit: Reservoirs and ecology of the nosocomial pathogen. J Infect Dis 168:943–947.

Hajjartabar M (2004) Poor-quality water in swimming pools associated with substantial risk of otitis externa due to *Pseudomonas aeruginosa*. Water Sci Technol 50:63–67.

Hall J, Callaway J, Tindall J, Smith J (1968) *Pseudomonas aeruginosa* in dermatology. Arch Dermatol 97:312–323.

Hambsch B, Sacre C, Wagner I (2004) Heterotrophic plate count and consumer's health under special consideration of water softeners. Int J Food Microbiol 92:365–373.

Hardalo C, Edberg SC (1997) *Pseudomonas aeruginosa*: Assessment of risk from drinking water. Clin Rev Microbiol 23:47–75.

Hentges DJ, Stein AJ, Casey SW, Que JU (1985) Protective role of intestinal flora against infection with *Pseudomonas aeruginosa* in mice: Influence of antibiotics on colonization resistance. Infect Immun 47:118–122.

Hewitt DJ, Weeks DA, Millner, Huss RG (2006) Industrial *Pseudomonas* folliculitis. Am J Ind Med 49:895–899.

Highsmith AK, Abshire RL (1975) Evaluation of a most-probable-number technique for the enumeration of *Pseudomonas aeruginosa*. Appl Microbiol 30:596–601.

Highsmith AK, Le PN, Khabbaz RF, Munn VP (1985) Characteristics of *Pseudomonas aeruginosa* isolated form whirlpools and bathers. Infect Control 6:407–412.

Hoadley AW (1977) Potential health hazards associated with *Pseudomonnas aeruginosa* in water. In: Hoadley AW, Dutka BJ (eds) Bacterial indicators/health hazards associated with water. American Society of Testing Materials, Philadelphia, PA, pp. 80–114.

Holder I (1993) *Pseudomonas aeruginosa* virulence-associated factors and their role in burn wound infections. In: Fick R (ed) *Pseudomonas aeruginosa* – The opportunist pathogenesis and disease. CRC, Boston, MA, p. 242.

Hollyoak V, Boyd P, Freman R (1994) Whirlpool baths in nursing homes: Use, maintenance, and contamination with *Pseudomonas aeruginosa*. Commun Dis Rep CDR Rev 5:R102–R104.

Hollyoak V, Allison D, Summers J (1995) *Pseudomonas aeruginosa* wound infection associated with a nursing home's whirlpool bath. Commun Dis Rep CDR Rev 5:R100–R102.

Hopkins RS, Abbott DO, Wallace LE (1981) Follicular dermatitis outbreak caused by *Pseudomonas aeruginosa* associated with a motel's indoor swimming pool. Public Health Rep 96:246–249.

Huang HI, Shih HY, Lee CM, Yang TC, Lay JJ, Lin YE (2008) In vitro efficacy of copper and silver ions in eradicating *Pseudomonas aeruginosa, Stenotrophomonas maltophilat*, and *Acinetobacter baumannii*: Implications for on-site disinfection for hospital infection control. Water Res 42:73–80.

Hudson PJ, Vogt RL, Jillson DA (1985) Duration of whirlpool use as a risk factor for *Pseudomonas* dermatitis. Am J Epidemiol 122:915–917.

Hunter CA, Ensign PR (1947) An epidemic of diarrhea in a new-born nursery caused by *Pseudomonas aeruginosa*. Am J Public Health 37:1166–1169.

Insler MS, Gore H (1986) *Pseudomonas* keratitis and folliculitis from whirlpool exposure. Am J Ophthalmol 101:41–43.

Jacobs RF, McCarthy RE, Elser JM (1989) *Pseudomonas osteochondritis* complicating puncture wounds of the foot in children: A 10-year evaluation. J Infect Dis 160:657–661.

Jacobson JA (1985) Pool-associated *Pseudomonas aeruginosa* dermatitis and other bathing-associated infections. Infect Control 6:398–401.

Kersters K, Ludwig W, Vancanneyt M, De Vos P, Gillis M, Schleifer K-H (1996) Recent changes in the classification of the pseudomonads: An overview. Syst Appl Microbiol 19:465–477.

Khabbaz RF, McKinley TW, Goodman RA, Hightower AW, Highsmith AK, Tait KA, Band JD (1983) *Pseudomonas aeruginosa* serotype 0:9 new cause of whirlpool-associated dermatitis. Am J Med 74:73–77.

Kiehn TE (1989) Bacteremia and fungemia in the immunocompromised patient. Eur J Clin Microbiol Infect Dis 8:832–837.

Kielhofner M, Atmar R, Hamill R, Musher D (1992) Life-threatening *Pseudomonas aeruginosa* infections in patients with human immunodeficiency virus infection. Clin Infect Dis 14:403–411.

Kolter R, Losick R (1998) One for all and all for one. Science 280:226–227.

Kominos SD, Copeland CE, Grosiak B, Postic B (1972) Introduction of *Pseudomonas aeruginosa* into a hospital via vegetables. Appl Microbiol 24:567–570.

Kreger BE, Craven DE, Carling PC, McCabe WR (1980) Gram-negative bacteremia. III. Reassessment of etiology, epidemiology and ecology in 612 patients. Am J Med 68:332–343.

Kush BJ, Hoadley AW (1980) A preliminary survey of the association of *Pseudomonas aeruginosa* with commercial whirlpool bath waters. Am J Pub Health 70:279–281.

Lermi A, Cunha BA (1998) *Pseudomonas aeruginosa* arterial line infection. Am J Infect Control 26:538.

Levesque B, Simard P, Gauvin D (1994) Comparison of the microbiological quality of water coolers and that of municipal water systems. Appl Environ Microbiol 60:1174–1178.

Losonsky GA, Hasan JAK, Huq A (1994) Serum antibody responses of divers to waterborne pathogens. Clin Diag Lab Immunol 1:182–185.

Manaia CM, Nunes OC, Morais PV, da Costa MS (1990) Heterotrophic plate counts and the isolation of bacteria from mineral waters on selective and enrichment media. J Appl Bacteriol 69:871–876.

Marrie TJ (1994) Community-acquired pneumonia. Clin Infect Dis 18:501–515.

Martins MT, Sato MIZ, Alves MN, Stoppe NC, Prado VM, Sanchez PS (1995) Assessment of microbiologic quality for swimming pools in South America. Water Res 29:2417–2420.

McClausland, WJ, Cox PJ (1975) Pseudomonas infection traced to motel whirlpool. J Environ Health 37:455–459.

McCubbin M, Fick RB (1992) Pathogenesis of *Pseudomonas* lung disease in cystic fibrosis. In: Fick R (ed) *Pseudomonas aeruginosa* – The opportunist pathogenesis and disease. CRC, Boston, MA, pp. 189–192.

McGowan JE, Barnes MW, Finland M (1975) Bacteremia at Boston City Hospital: Occurrence and mortality during 12 selected years (1935–1972), with special reference to hospital acquired cases. J Infect Dis 132:316–335.

McManus AT, Mason AD, McManus WF, Pruitt BA (1985) Twenty-five year review of *Pseudomonas aeruginosa* bacteremia in a burn center. Eur J Clin Microbiol 4:219–223.

Mendelson M, Gurtman A, Szabo S, (1994) *Pseudomonas aeruginosa* bacteremia in patients with AIDS. Clin Infect Dis 18:886–895.

MMWR (1979) Rash associated with use of whirlpools – Maine. Morb Mortal Wkly Rep 28:182–184.

MMWR (1983) An outbreak of *Pseudomonas* folliculitis associated with a waterslide – Utah. Morb Mortal Wkly Rep 32:425–427.

Molina DN, Colon M, Bermudez RH, Ramirez-Ronda CH (1991) Unusual presentation of *Pseudomonas aeruginosa* infections: A review. Bol Asoc Med P R 83:160–163.

Moreira L, Agostinho P, Morals PV, da Costa MS (1994) Survival of allochthonous bacteria in still mineral water bottled in polyvinyl chloride (PVC) and glass. J Appl Bacteriol 77:334–339.

Muyldermans G, de Smet F, Pierard D, Steenssens L, Stvens D, Bougatef A, Lauwers S (1998) Neonatal infections with *Pseudomonas aeruginosa* associated with a water-bath used to thaw fresh frozen plasma. J Hosp Infect 39:309–314.

Noble RC, Overman SB (1994) Pseudomonas stutzeri infection. A review of the literature. Diagn Microbiol Infect Dis 19:51–56.

Ollstein R, McDonald C (1980) Topical and systemic antimicrobial agents in burns. Ann Plast Surg 5:386–392.

Papapetropoulou M, Iliopoulou J, Rodopoulou G (1994) Occurrence of antibiotic-resistance of *Pseudomonas* species isolated from drinking water in southern Greece. J Chemother 6:111–116.

Payment P, Gamache F, Paquette G (1988) Microbiological and virological analysis of water from two water filtration plants and their distribution systems. Can J Microbiol 34:1304–1309.

Pellett S, Bigley DV, Grimes DJ (1983) Distribution of *Pseudomonas aeruginosa* in a riverine ecosystem. Appl Environ Microbiol 45:328–332.

Pollack M (1995) *Pseudomonas aeruginosa*. In: Mandell GL, Bennett JE, Colin R (eds) Principles and practice of infectious diseases, 4th Ed. Churchill Livingstone, New York, NY, pp. 1980–2003.

Porco FV, Visconte EB (1995) *Pseudomonas aeruginosa* as a cause of infectious diarrhea successfully treated with oral ciprofloxacin. Ann Pharmacother 29:1122–1123.

Price D, Ahearn D (1988) Incidence and persistence of *Pseudomonas aeruginosa* in whirlpools. J Clin Microbiol 26:1650–1654.

Pyle B, McFeters G (1989) Iodine sensitivity of bacteria isolated from iodinated water systems. Can J Microbiol 35:520–523.

Ramphal R, McNiece M, Pollack F (1981) Adherence of *Pseudomonas aeruginosa* to the injured cornea: A step in the pathogenesis of corneal infections. Ann Ophthalmol 13:421–425.

Ratnam S, Hogan K, March SB, Butler RW (1986) Whirlpool-associated folliculitis caused by *Pseudomonas aeruginosa*: Report of an outbreak and review. J Clin Microbiol 23:655–659.

Reid TMS, Porter IA (1981) An outbreak of otitis externa in competitive swimmers due to *Pseudomonas aeruginosa*. J Hyg Camb 86:357–362.

Reisler RB, Blumberg H (1999) Community-acquired *Pseudomonas stutzeri* vertebral osteomyelitis in a previously healthy patient: Case report and review. Clin Infect Dis 29:667–669.

Reitler R, Seligmann R (1957) *Pseudomonas aeruginosa* in drinking water. J Appl Bacteriol 20:145–150.

Relai D, Rosati S (1994) Antibiotic susceptibility and serotyping of *Pseudomonas aeruginosa* strains isolated form surface waters, thermomineral waters and clinical specimens. Zbl Hyg 196:75–80.

Remington J, Schimpff S (1981) Please don't eat the salads. N Eng J Med 304:433–435.

Restaino L, Frampton EW, Hemphill JB, Palnikar P (1995) Efficacy of ozonated water against various food-related microorganisms. Appl Environ Microbiol 61:3471–3475.

Roberts LA, Collignon PJ, Cramp VB, Alexander S, McFarlane AE, Graham E, Fuller A, Sinickas V, Hellyar A (1990) An Australia-wide epidemic of *Pseudomonas picketti* bacteraemia due to contaminated "sterile" water for injection. Med J Aust 152:652–655.

Rogues AM, Boulestreau H, Lasheras A, Boyer A, Gruson D, Merle C, Casting Y, Bebear CM, Gachie JP (2007) Contribution of tap water to patient colonization with *Pseudomonas aeruginosa* in a medical intensive care unit. J Hosp Infect 67:72–78.

Rose HD, Franson TR, Sheth NK, (1983) *Pseudomonas* pneumonia associated with use of a home whirlpool spa. JAMA 250:2027–2031.

Rusin PA, Rose JB, Haas CN, Gerba CP (1997) Risk assessment of opportunistic bacterial pathogens in drinking water. Rev Environ Contam Toxicol 152:57–83.

Rusin P, Orosz-Coughlin P, Gerba C (1998) Reduction of faecal coliform, coliform and heterotrophic plate count bacteria in the household kitchen and bathroom by disinfection with hypochlorite cleaners. J Appl Microbiol 85:819–828.

Salahuddin N, Madhavan T, Fisher E (1973) *Pseudomonas* osteomyelitis. Radiologic features. Radiology 109:41–47.

Salmen P, Dwyer DM, Vorse H, Kruse W (1983) Whirlpool-associated *Pseudomonas aeruginosa* urinary tract infections. JAMA 250:2025–2026.

Saroff A, Armstrong D, Johnson W (1973) *Pseudomonas* endocarditis. Am J Cardiol 32:234–237.

Sausker WF (1978) *Pseudomonas* folliculitis acquired from a health spa whirlpool. JAMA 239:2362–2365.

Schillinger J, Du Vall Knorr, S (2004) Drinking-water quality and issues associated with water vending machines in the city of Los Angeles. J Environ Health 66:25–31.

Schimpff SC, Wiernik PH, Block JB (1972) Rectal abscesses in cancer patients. Lancet 2:844–847.

Seyfried PL, Cook RJ (1984) Otitis externa infections related to *Pseudomonas aeruginosa* levels in five Ontario lakes. Can J Public Health 75:83–91.

Shehabi AA, Masoud H, Maslamani FA (2005) Common antimicrobial resistance patterns, biotypes and serotypes found among *Pseudomonas aeruginosa* from patient's stool and drinking water sources in Jordan. J Chemother 17:179–183.

Shekar R, Rice T, Zierdt C, Kallick C (1985) Outbreak of endocarditis caused by *Pseudomonas aeruginosa* serotype 011 among pentazocine and tripelnnamine abusers in Chicago. J Infect Dis 151:203–208.

Shelley AW, Deeth HC, MacRae IC (1987) A numerical taxonomy study of psychrotrophic bacteria associated with lipolytic spoilage of raw milk. J Appl Bacteriol 62:197–207.

Shigemura K, Arakawa S, Sakai Y, Kinoshita S, Tanaka K, Fujisawa M (2006) Complicated urinary tract infection caused by *Pseudomonas aeruginosa* in a single institution (1999–2003). Int J Urol 13:538–542.

Shooter RA (1971) Bowel colonization of hospital patients by *Pseudomonas aeruginosa* and *Escherichia coli*. Proc R Soc Med 64:27–28.

Shooter RA, Gaya H, Cooke EM, Kumar P, Patel N (1969) Food and medicaments as possible sources of hospital strains of *Pseudomonas aeruginosa*. Lancet 1:1227–1229.

Silvestry-Rodriquez N, Sicairos-Ruelas AE, Gerba CP, Bright KR (2007a) Silver as a disinfectant. Rev Environ Contam Toxicol 191:23–45.

Silvestry-Rodriquez N, Bright KR, Uhlmann DR, Slack DC, Gerba CP (2007b) Inactivation of *Pseudomonas aeruginosa* and *Aeromonas hydrophila* by silver in tap water. J Environ Sci Health 42:1579–1584.

Solomon SL (1985) Host factors in whirlpool-associated *Pseudomonas aeruginosa* skin disease. Infect Control 6:402–405.

Springer GL, Shapiro ED (1985) Fresh water swimming as a risk factor for otitis externa: A case–control study. Arch Environ Health 40:202–206.

Stone HH, Kolb LD, Geheber CE (1979) Bacteriologic considerations in perforated necrotizing enterocolitis. South Med J 72:1540–1543.

Sundheim, Sletten GA, Dainty RH (1998) Identification of pseudomonads from fresh and chill-stored chicken carcasses. Int J Food Microbiol 39:185–194.
Tamagnini LM, Gonzalez RD (1997) Bacteriological stability and growth kinetics of *Pseudomonas aeruginosa* in bottled water. J Appl Microbiol 83:91–94.
Tancrede CH, Andremont AO (1985) Bacterial translocation and gram-negative bactremia in patients with hematological malignancies. J Infect Dis 152:99–103.
Thomas P, Moore M, Bell E (1985) *Pseudomonas* dermatitis associated with a swimming pool. JAMA 253:1156–1159.
Torres A, Aznar R, Gatell JM (1990) Incidence, risk, and prognosis factors of nosocomial pneumonia in mechanically ventilated patients. Am Rev Respir Dis 142:523–528.
Torres A, Serra-Batelles J, Ferrer A (1991) Severe community-acquired pneumonia. Am Rev Respir Dis 144:312–318.
Traill ZC, Miller RF, Ali N, Shaw PJ (1996) *Pseudomonas aeruginosa* bronchopulmonary infection in patients with advanced human immunodeficiency virus disease. Br J Radiol 69:1099–1103.
Trautmann M, Leppper PM, Haller M (2005) Ecology of *Pseudomonas aeruginosa* in the intensive care unit and the evolving role of water outlets as a reservoir of the organism. Am J Infect Control 33:S41–S49.
Trautmann M, Halder S, Josef H, Hilde R Haller M (2008) Point-of-use water filtration reduces endemic *Pseudomonas aeruginosa* infections on a surgical intensive care unit. Am J Infect Control 36:421–429.
Tredget EE, Shankowsky HA, Joffe AM, Inkson TI, Volpel K, Paranchych W, Kibsey PC, Alton JD, Burke JF (1992) Epidemiology of infections with *Pseudomonas aeruginosa* in burn patients: The role of hydrotherapy. Clin Infect Dis 15:941–949.
van Asperen IA, de Rover CM, Schijven JF (1995) Risk of otitis externa after swimming in recreational fresh water lakes containing *Pseudomonas aeruginosa*. Br Med J 311:407–410.
van der Kooij D, Oranje JP, Hijnen WAM (1982) Growth of *Pseudomonas aeruginosa* in tap water in relation to utilization of substrates at concentrations of a few micrograms per liter. Appl Environ Microbiol 44:1086–1095.
Vanholder R, Vanhaecke E, Ringoir S (1990) Waterborne *Pseudomonas septicemia*. ASAIO Trans 36:M215–M216.
Vess R, Anderson R, Carr J (1993) The colonization of solid PVC surfaces and the acquisition of resistance to germicides by water microorganisms. J Appl Bacteriol 74:215–221.
Vianelli N, Giannini MB, Quarti C, Sabattini MAB, Fiacchini M, de Vivo A, Graldi P, Galli S, Nanetti, Baccarani M, Ricci P (2006) Resolution of *Pseudomonas aeruginosa* outbreak in a hematology unit with the use of disposable sterile water filters. Haematologica 91:983–985.
Warburton DW, Dodds KL, Burke R (1992) A review of the microbiological quality of bottled water sold in Canada between 1981 and 1989. Can J Microbiol 38:12–19.
Warburton DW, Bowen B, Konkle A (1994) The survival and recovery of *Pseudomonas aeruginosa* and its effect upon *Salmonellae* in water: Methodology to test bottled water in Canada. Can J Microbiol 40:987–992.
Ward NR, Wolfe RL, Olson BH (1984) Effect of pH, application technique, and chlorine-to-nitrogen ratio on disinfectant activity of inorganic chloramines with pure culture bacteria. Appl Environ Microbiol 48:508–514.
Washburn J, Jacobson JA, Marston E, Thorsen B (1976) *Pseudomonas aeruginosa* rash associated with a whirlpool. JAMA 235:2205–2207.
Weber G, Werner HP, Matshnigg H (1971) *Pseudomonas aeruginosa* im Trinkwaqsser als Todesursache bei Neugeborenen. Zbl Bakt I Abt Orig 216:210–214.
Whimbey E, Gold JWM, Polsky B (1986) Bacteremia and fungemia in patients with the acquired immunodeficiency syndrome. Ann Intern Med 104:511–514.
Wiesseman G, Wood V, Kroll L (1973) *Pseudomonas* vertebral osteomyelitis in heroin addicts. Report of five cases. J Bone Joint Surg 55:1416–1424.

Williams DE, Worley SD, Wheatley WB, Swango LJ (1985) Bactericidal properties of a new water disinfectant. Appl Environ Microbiol 49:637–643.

Witt DJ, Craven DE, McCabe WR (1987) Bacterial infections in adult patients with the acquired immune deficiency syndrome (AIDS) and AIDS-related complex. Am J Med 82:900–906.

Wolfe RL (1990) Ultraviolet disinfection of potable water. Environ Sci Technol 24:768–773.

Zichichi L, Asta G, Noto G (2000) *Pseudomonas aeruginosa* follicultis after shower/bath exposure. Int J Dematol 39:270–273.

Ziegler EJ, Fisher CJ, Sprung CL (1991) Treatment of gram-negative bacteremia and septic shock with HA-1A human monoclonal antibody against toxin. N Eng J Med 324:429–436.

Non-thermal Plasmas Chemistry as a Tool for Environmental Pollutants Abatement

Yan-hong Bai, Jie-rong Chen, Xiao-yong Li, and Chun-hong Zhang

Contents

1	Introduction	117
2	Generation and Chemical Mechanisms of NTPs Chemistry	119
	2.1 Generation	119
	2.2 Chemical Mechanisms	119
3	Applications of NTPs Chemistry	121
	3.1 Removal of Atmospheric and Gaseous Pollutants	121
	3.2 NTPs Chemistry for Wastewater Treatment	125
	3.3 NTPs Chemistry for Quality of Life and Safety	127
	3.4 NTPs Chemistry for Solid Waste Disposal	129
4	Future Trends	129
5	Summary	130
References		130

1 Introduction

Increasing concern for protecting the environment against the effects of growing industrialization, intensive agriculture, and further exploitation of natural resources poses enormous challenges to contemporary science and to the sustainable development of mankind. Most environmental pollutants pose some harm to human health and the environment irrespective of whether exposures occur through air, water, soil, or the food chain. Therefore, innovative alternative technologies and equipment are needed to abate, convert, decompose, or otherwise manage environ-

Y.-h. Bai (✉)
School of Life Science and Technology, Xi'an Jiaotong University,
Xi'an, 710049, People's Republic of China
Department of Chemistry, School of Science, Xi'an Jiaotong University,
Xi'an 710049, People's of China
email: yhbai7@mail.xjtu.edu.cn

J.-r. Chen (✉)
Department of Environmental Science and Engineering, Xi'an Jiaotong University,
Xi'an 710049, People's Republic of China
e-mail: jrchen@mail.xjtu.edu.cn

mental pollutants. Non-thermal plasma (NTPs) chemistry is an innovative tool for potentially abating environmental pollutants; this technology promises to meet a host of abatement demands, including those mentioned above, and also offers crucial advantages such as improved energy efficiency, higher levels of treatment in smaller spaces, and near zero-emissions.

The origin of NTPs chemistry can be traced to experiments conducted in 1857 (Siemens 1857); this work addressed the silent discharge (also referred to as dielectric-barrier discharge (DBD)) for generating ozone. At that time, ozone was primarily used to disinfect drinking water. Since this early period, most NTPs chemical studies have involved organic film synthesis and preparation, in which molecules such as C_2H_2 and NH_3 play a role. Tonks and Langmuir (1929) first used the word plasma to designate that portion of an arc-type discharge, in which densities of ions and electrons are high, but are substantially equal. The term "plasma chemistry" was first used by McTaggart (1967) in his book named "Plasma Chemistry in Electrical Discharges." Plasma constitutes a partially or fully ionized gas consisting of various particles such as electrons, free radicals, ions, atoms, and molecules. There are two categories of plasmas (1) thermal plasmas and (2) non-thermal plasmas. In thermal plasma, sufficient energy is introduced to allow plasma constituents to be in thermal equilibrium, i.e., the ions and electrons are, on average, at the same temperature. The temperature of thermal plasma components is about 1–2 eV (1 eV is associated with 11,600 K). The focus of this review is on non-thermal plasmas, rather than thermal plasmas. A NTPs is one in which the mean electron energy, or temperature, is considerably higher than that of the bulk-gas molecules. Because energy is added to the electrons instead of to the ions and the background gas molecules, the electrons can attain energies of 1–10 eV, whereas the background gas remains at ambient temperature. Therefore, NTPs are also referred to as nonequilibrium or "cold" plasmas. This non-thermal condition differs from that of common gases, in that NTPs are good sources of highly reactive oxidative and reductive species and plasma electrons that are capable of reacting with and decomposing chemical pollutants to yield safer products. As a result, NTPs chemistry may solve many pollutant abatement problems that traditional methods cannot easily address.

In recent decades, there has been considerable interest in applying NTPs to address problems such as thin-film deposition and etching in the microelectronics industries (Alexandrov and Hitchman 2005; Petrovic and Makabe 1998), various surface treatment applications (Deng et al. 2007; Li and Chen 2006; Laguardia et al. 2005; Chu et al. 2002), and chemical synthesis (Nair et al. 2007; Dyuzhev and Ioffe 2002). NTPs have also shown good performance in degrading low concentrations of toxic compounds. Therefore, NTPs chemistry has also been employed experimentally, and in commercial field studies, to destroy environmental pollutants including removal of volatile organic compounds (VOCs), odors from air, cleaning diesel exhaust, wastewater treatment, and sterilization of medical or packaging equipment. Although a detailed understanding of how NTPs chemistry works is still unavailable, we wish to address known aspects that pertain to it in this review. Specifically, the purpose of this paper is to provide a basic understanding of what is known of NTPs chemistry as it relates to pollution abatement. The emphasis on NTPs chemistry in this review will be on chemical mechanisms, useful applications, and the future of NTPs chemistry for providing solutions in the field of environmental pollutants abatement.

2 Generation and Chemical Mechanisms of NTPs Chemistry

2.1 Generation

There are several ways in which NTPs are generated for use in applications for environmental pollutants abatement. Such approaches include glow discharges, pulsed corona discharges, dielectric barrier discharges, radio frequency discharges, and microwave discharges. The performance of NTPs can be significantly enhanced by altering generating parameters such as electrode shape, the nature and strength of the power supply, plasma gas and flow rates, and the concentration of pollutants that are being treated. Energy efficiency and treatment selectivity are improved when more capable plasmas, that have the proper combination of catalyst reactors, are employed. Several review papers have been written on how such parameters can affect NTPs processing (Mizuno 2007; Louis 2005; Kim 2004; Schutze et al. 1998; Eliasson and Kogelschatz 1991a).

2.2 Chemical Mechanisms

Although large differences in methods exist for inducing chemical reactions, the reactions themselves are similar. NTPs are unique in inducing various nonequilibrium chemical reactions at room temperature. Highly energetic electrons are the entities by which NTPs chemicals can effectively degrade environmental pollutants. Such electrons act as "electronic scissors," and can provide sufficient energy to break the chemical bonding of virtually any gas molecule by inelastic collision (Masuda 1988) with background molecules; these reactions also produce secondary electrons and highly reactive species (free-radicals, ions, and excited molecules, such as ·OH, ·O radicals, and ozone) by ionization, excitation, and dissociation (Savinov et al. 2003). It is these intermediate species that promote the desired conversions of pollutants at uncharacteristic low temperatures.

The energy in NTPs chemistry electrons typically ranges from 1 to 10 eV; this energy range is ideal for excitation of atomic and molecular species and for breaking chemical bonds (Eliasson and Kogelschatz 1991b). In the excitation process, if one reactant is brought to an excited state, it can surmount the active energy E_a and thus initiate the chemical reaction. Generally, the E_a of most chemical reactions is <5 eV (McQuarrie and Rock 1984). Alternatively, if one wants to dissociate a molecule to form a free radical, the energy delivered to that molecule must correspond to the strength of the chemical bond required to break it. The average bond dissociation energies of some diatomic molecules (most covalent bond energy equals 3–6 eV), which relate to the degree of difficulty for breaking bonds, are shown in

Table 1 Energy associated with non-thermal plasma (NTPs) activated particles, and examples of bond energies (298.15 K, 100 kPa)

Activated particles	Energy (eV)	Bond	Bond energy (eV)	Bond	Bond energy (eV)	Bond	Bond energy (eV)
Electrons	1–10	C–H	4.31	C–N	2.88	S–H	3.83
Ions	0–2	C–C	3.61	C–P	2.74	S–O	5.43
Excited particles	0–20	C–Cl	3.44	C=O	7.76	S–C	2.66
Photons	3–40	C–Br	2.87	N–O	1.83	S–P	2.40
		C–O	3.66	P–O	5.23	S=O	4.78

Table 1 (Chang 2002). In addition, some ionic reaction pathways of decomposition are peculiar to the structural features of specific VOCs (Marotta et al. 2005).

At present, it is generally accepted that the mechanisms by which environmental pollutants are degraded through use of NTPs are based on chain cycles via radicals, mainly ˙OH, ˙O, ˙H radicals, and ozone. All of these radicals are strongly reactive species, with high reduction potential, i.e., 2.42 V for ˙O, 2.07 V for O_3, 2.80 V for ˙OH, and 1.78 V for H_2O_2. When pollutant X reacts with an electron, one or more of the following species are generated: ˙OH, ˙O, ˙H, and O_3. Such species produce a chain reaction (radical-promoted) that liberates final products that are less hazardous than were the original pollutants. Environmental pollutant X may be inorganic (SO_2, NO_2, H_2S, NH_3, CO_2, CO, etc.), or organic (hydrocarbons, halohydrocarbons, alcohols, phenols, aldehydes, ketones, etc.). The major decomposition chemical mechanisms and reaction steps that are initiated by NTPs reactive species are presented below (Mizuno 2007; Savinov et al. 2003; Ono and Oda 2000; Penetrante et al. 1997; Yamamoto et al. 2003; Lovejoy et al. 1990):

$$e + O_2 \rightarrow \text{˙O} + \text{˙O} \quad (1)$$

$$\text{˙O} + O_2 \rightarrow O_3 \quad (2)$$

$$e + H_2O \rightarrow \text{˙OH} + \text{˙H} \quad (3)$$

$$\text{˙OH} + \text{˙OH} \rightarrow H_2O_2 \quad (4)$$

$$e + X \rightarrow \text{˙X} \quad (5)$$

$$\text{˙X} + X \rightarrow \text{products} \quad (6)$$

$$\text{˙O}, \text{˙OH}, O_3, H_2O_2 + X \rightarrow \text{products} \quad (7)$$

In short, the removal efficiency of a pollutant by NTPs depends mainly on:

1. The ability to produce large amounts of ˙O, ˙OH radicals, and O_3 in the reaction, and
2. The rates of reactions of the pollutant molecules with the above reactive species.

Therefore, it is crucial to clarify all relevant chemical pathways for optimizing NTPs processes when novel applications are sought in environmental pollutants abatement. In particular, identification of the various possible products/reactive species that are formed during NTPs processing is needed. Another requirement is the diagnosis and tracking of reactive species, mainly detection of ˙OH radicals and other reactive radicals in the NTPs, by using laser-induced fluorescence (LIF) technology (Ono and Oda 2000; Magne and Pasquiers 2005). Such analysis is very important in understanding what happens in NTPs and how to improve the degradation efficiency of various pollutants, compared with classical chemical reactions. Generally, the following qualitative and quantitative analytical techniques can be useful in tracing kinds and levels of residual pollutants and byproducts during decomposition of pollutants through applications of NTPs chemistry: luminescence detection, infrared absorption, UV absorption, gas chromatography-mass spectrometry (GC-MS), and high pressure liquid chromatography-mass spectrometry (HPLC-MS).

3 Applications of NTPs Chemistry

3.1 *Removal of Atmospheric and Gaseous Pollutants*

Air and gaseous pollutants that occur as atmospheric aerosols such as SO_2, NO_x, H_2S, NH_3, CO_2, PFCs (perfluoro-carbons), formaldehyde, VOCs, radioactive gases, etc. are routinely released from polluting industrial sources, automotive traffic, and as byproducts from use of other chemicals. Such air pollutants are urgent topics for environmental pollution abatement research. These pollutants are important because of their environmental and ecological impact; such impacts include atmospheric acidification, smog, induction of the greenhouse effect, global warming, ozone depletion, and potential induction of environmental toxicity. As investigators have searched for alternative abatement technologies to address such pollutants over the last decade, many have suggested that NTPs could be one of the most effective methods for pollutant removal. Two main types of NTPs reactors may be used to treat pollutants (1) pulsed corona streamer and (2) silent (barrier) discharge reactors. Both reactor types have potential advantages for control of air and gaseous pollutants. To enhance the effectiveness of treating gaseous or atmospheric pollutants, catalysts may be used. If they are used, then the process is referred to as plasma-assisted catalysis.

3.1.1 Removal of NO_x and SO_2

Oxidation and reduction are the two primary ways in which NTPs are applied to simultaneously remove NO_x and SO_2 and other toxic gaseous pollutants (Ken et al. 2007; Khacef and Cormier 2006; Khacef et al. 2002; Kuroki et al. 2002; Yan et al. 1999; Higashi et al. 1992; Clements et al. 1989). In the oxidation pathway, NO_x and SO_2 pollutants are simultaneously oxidized by reactive radicals to form NO_2 and SO_2, respectively, using wet discharge plasma reactors; this is followed by neutralization by NH_3 to produce solid ammonium sulfate and ammonium nitrate, which may be used as fertilizers (Clements et al. 1989; Peter et al. 2007; Mizuno 2000).

It has been demonstrated that ·OH and ·O radicals are the keys to oxidation of NO_x and SO_2 and act by chain cycle reactions (Khacef and Cormier 2006; Kuroki et al. 2002). Alternatively, NO_x and SO_2 can be reduced to form S, N_2, and O_2, when their chemical bonds are broken through energy provided by NTPs (Kim et al. 2001; Eichwald et al. 1997; Penetrante et al. 1995); such reductive processes take place in vehicle exhausts. Accordingly, the overall chemical mechanism for SO_2 and NO_x conversions is summarized as follows:

$$\text{Oxidation}: \cdot OH + SO_2 \rightarrow HSO_3 \quad (8)$$

$$\cdot OH + HSO_3 \rightarrow H_2SO_4 \quad (9)$$

$$\cdot OH + NO_2 \rightarrow HNO_3 \quad (10)$$

$$\cdot OH + NO \rightarrow HNO_2 \quad (11)$$

$$2NH_3 + H_2SO_4 \rightarrow 2(NH_4)_2SO_4 \quad (12)$$

$$NH_3 + HNO_3 \rightarrow NH_4NO_3 \quad (13)$$

$$\text{Reduction}: \cdot O + NO \rightarrow NO_2 \quad (14)$$

$$NO + NO_2 + 2NH_3 \rightarrow 2N_2 + 3H_2O \quad (15)$$

Because of the extremely high power consumption of NTPs for decomposing NO_x and SO_2, both in oxidation and reduction processes, researchers have sought catalysts that can work with NTPs-compatible support materials. Such catalysts and additives may include ammonia (Ken et al. 2007), water vapor, and hydrocarbons (Ingelsten et al. 2004; Oda et al. 1998), and γ-Al_2O_3 (Yoshihiko et al. 2006), which are usually combined with the NTPs process to enhance decomposition efficiency both in academic studies and for use in industrial applications. It has been reported in the literature that addition of hydrocarbons (mainly C_3H_6, C_2H_6, and CH_4) is known to drastically enhance NO_2 reduction; hydrocarbons are preferable to NH_3 as catalysts when NTPs treatment is used in vehicle exhausts (Li et al. 2007). In addition, Yamamoto et al. (2003) has reported the performance of NTPs reactors when used to oxidize NO in diesel engine exhaust gas. These authors demonstrated that NO was oxidized to NO_2 and then totally converted to N_2 and Na_2SO_4, using Na_2SO_3 as the catalyst.

3.1.2 Removal of Odors

Environmental odors are increasingly of concern to people, whether or not they are potentially hazardous. Agents such as NH_3, H_2S, and CS_2 are among those odorous gases of most concern. They are emitted from factories, exist in indoor air, or are emitted from facilities that house animals. These substances are of concern not only because they smell bad, but also because they contribute to deterioration of air quality and may affect human health. NTPs chemistry may be used to reduce odors without causing secondary pollution, as does some treatment methods (Jarrige and Vervisch 2007; Shi et al. 2006; Nicole et al. 2005; Ma et al. 2001; Chang and Tseng 1996; Lovejoy et al. 1990; Fateev et al. 2005). NTPs chemistry is more than 95% effective in removing odors; the decomposition pathways of these odors are dominated by ·OH, ·O, and O_3 via cogent radical attack products, thus producing major products such as H_2O, SO_2, CO_2, CO, and NH_4NO_3. All of these are less odorous than the original compounds.

The radical-driven decomposition process for H_2S is based on chain reaction cycles via ·OH and ·O radicals, as shown in (16)–(19):

$$H_2S + ·O \rightarrow HS· + ·OH \qquad (16)$$

$$H_2S + ·OH \rightarrow HS· + H_2O \qquad (17)$$

$$HS· + O_2 \rightarrow SO· + ·OH \qquad (18)$$

$$SO· + ·O \rightarrow SO_2 \qquad (19)$$

For removal of ammonia, NH_3 is reacted with hydroxyl radicals to produce N_2 through the NH_2 radical intermediates. The reactions are shown in (20)–(22):

$$NH_3 + ·OH \rightarrow H_2O + ·NH_2 \qquad (20)$$

$$NH_3 + 4O· \rightarrow NO· + 3·OH \qquad (21)$$

$$·NH_2 + NO· \rightarrow N_2 + H_2O \qquad (22)$$

One major concern associated with using NTPs to decompose H_2S and NH_3 is that it produces SO_2 and NO_x as secondary hazards. Fortunately, SO_2 and NO_x may be further decomposed through the free-radical NTP chemical reactions similar to those that produce SO_2 and NO_x.

Another odorous compound, CS_2, is first oxidized to form SO_2 and then other products are formed such as H_2SO_4. The reaction proceeds as shown in (23)–(25):

$$·OH + CS_2 \rightarrow CS_2OH \qquad (23)$$

$$CS_2OH + O_2 \rightarrow X \qquad (24)$$

$$X + O_2 \rightarrow SO_2 \rightarrow products \qquad (25)$$

3.1.3 Removal of VOCs

VOCs such as alkanes, alkenes, alcohols, aldehydes, ketones, and aromatic or halogenated compounds are hazardous pollutants emitted from paints, solvents, preservatives, automobile exhaust gas, and certain industrial processes. Some VOCs may be carcinogens or may produce respiratory diseases (Forst and Conroy 1998). At present, one of the most promising ways to degrade VOCs is through application of NTPs chemistry. NTPs chemistry has been used to destroy butane and benzene (Futamura et al. 1998, 2004), methyl chloride (Hsieh et al. 1998), toluene, acetone, isopropyl alcohol (Kogelschatz 2003), methanol and the other VOCs (Futamura and Sugasawa 2008), dichloromethane (Anna et al. 2007), trichlorethylene (Magureanu et al. 2007), bromomethane (Zhang et al. 1999), acetylene (Thevenet et al. 2007), and CFC-12 (Wallis et al. 2007). These reports show that NTPs have high efficiency (more than 90%) in destroying VOCs, when compared with other elimination methods. Various by-products result from the application of NTPs to destroy VOCs. These byproducts include the following: CO_2, H_2O_2, CO, NO_x, CClFO, $CHClF_2$, and C_2ClF_2. Some of these reaction products are themselves environmentally undesirable and result when NTPs decomposition is incomplete (Wallis et al. 2007; Penetrante et al. 1997). To reduce the levels of undesired byproducts during treatment of VOCs, NTPs are sometimes combined with an appropriate catalyst such as $BaTiO_3$, Al_2O_3, TiO_2, MnO_2, and zeolites or their derivatives (Kim et al. 2008; Wallis et al. 2007; Demidiouk and Chae 2005; Sano et al. 2006; Ogata et al. 2003; Oh et al. 2006). Among these choices of catalysts, TiO_2 is often revealed as having a higher effectiveness and better energy efficiency.

The chemical reaction mechanisms that result in decomposition of VOCs, when using NTPs chemistry, are often quite complicated. Various research groups have performed investigations to better understand the mechanisms of by-products formation and how VOCs are decomposed when NTPs are employed for VOCs destruction (Penetrante et al. 1997; Lee and Chang 2001; Futamura and Yamamoto 1997). It is generally accepted that the mechanism of VOCs destruction is dominated by free-radical reactions, both in dry and wet environments; wet environments result in better decomposition than dry ones because the concentrations of free radicals are usually higher in wet environments. Yamamoto (1997) has proposed the explanation that electron impact acts as the agent for transfer of electron energy during the degradation of VOCs. VOC molecules are therefore degraded when sufficient energy is accumulated to overcome potential energy barriers and radicals are formed; at this point VOCs are degraded homolitically as a result of their excited states. This energy transfer concept indicates that VOCs are degraded more easily when sufficient energy, derived from both electrons and reactive species, become available to break the VOC chemical bonds. In other studies, evidence indicates that ionic reactions constitute another pathway that is capable of decomposing VOCs by NTPs; examples of this pathway are decomposition of CH_3OH (Penetrante et al. 1997) and chlorinated VOCs (Marotta et al. 2005).

Notwithstanding the above investigations, the basic kinetics of VOCs destruction remains unclear. Some experiments, but not others, have shown that the kinetics of destruction strongly depends on the initial concentration and structure of the VOCs, as well as on the intensity of energy provided by NPTs. These differences of opinion can be resolved experimentally. Most results indicated that the degradation rate of VOCs follows first-order kinetics (Futamura and Yamamoto 1997; Rudolph et al. 2002; Mok et al. 2002; Futamura et al. 1997). However, others have proposed that reactions may follow second-order kinetics at certain concentrations of VOCs and reactive species derived from NTPs (Koutsospyros et al. 2005; Yan et al. 2001).

3.1.4 Purification of Indoor Air

Indoor air pollutants, including formaldehyde, toluene, the radioactive gas Radon, bacteria, and odorants, which are released from building and decorative materials and human activities (contact with printers and computers, tobacco smoking, cooking, solvent fumes, etc.), are of concern because they may cause adverse health effects (Jones 1999). There are a few studies on use of NTPs for removal of such indoor air pollutants (Van Durme et al. 2007; Chang 2003; Chang and Lee 1995; Storch and Kushner 1993). These studies invariably reported on DBD plasma as a means to remove formaldehyde, toluene, and acetaldehyde at destruction efficiencies exceeding 70%. The synergistic effect of combining catalysts (such as TiO_2 and MnO_2) and NTPs to enhance removal performance of indoor air pollutants was also studied (Subrahmanyam et al. 2007; Xu et al. 2003). The chemical pathways by which plasma-assisted catalysts operate are similar to those in atmospheric chemistry. Catalysts strongly accelerate the formation of ˙OH and ˙O radicals, and result in improved degradation efficiencies, such that they may reach nearly 100% for indoor pollutants like toluene or formaldehyde (HCHO).

The degradation mechanism for HCHO is as follows:

$$˙OH + HCHO \rightarrow CHO + H_2O \qquad (26)$$

$$CHO + ˙O \rightarrow CO_2 + H_2 \qquad (27)$$

It has been confirmed that NTPs chemistry is very effective for abatement of low concentrations of gaseous pollutants. Moreover, NTPs chemistry has lower investment and operational costs, and multiple gaseous pollutants can be removed simultaneously. However, further commercialization of the technology will be limited by the relatively low-energy conversion efficiency and the fact that the chemical pathway is not yet clear. Thus, different approaches are needed to further develop and refine applications of NTPs for indoor air pollutants. To clarify the mechanisms by which chemical indoor air pollutants are destroyed, future reaction diagnoses are needed for active species generated in NTPs by LIF or other fast response optical technologies.

3.2 NTPs Chemistry for Wastewater Treatment

In recent years, NTPs chemistry has received a great deal of attention as a potential application for wastewater treatment (Locke et al. 2006; Lukes et al. 2005; Sahni et al. 2005; Hao et al. 2007). Both pulsed corona discharge and DBD (either directly in water or in the gas phase above the water) have been demonstrated to produce chemically active species, such as ˙OH, ˙H, and ˙O radicals, and O_3, H_2O_2 molecules (Kirkpatrick and Locke 2005; Lukes et al. 2004; Šunka et al. 1999). In addition, UV light, generated from NTPs in water (Šunka 2001), may also promote production of superoxides or peroxides (Yen et al. 2000). UV light and chemical reactive species, especially hydroxyl radicals, have been shown to rapidly and efficiently degrade many organic compounds, including phenol (Lukes et al. 2005; Joshi et al. 1995), 2,4-dinitrophenol (Zhang et al. 2008), aniline (Tezuka and Iwasaki 2001),

4-chlorophenol, 2,4,6-trinitrotoluene, 4-dichloroaniline (Hao et al. 2007; Willberg et al. 1996), trichloroethylene (Sahni et al. 2005), acetophenone (Wen and Jiang 2001), and organic dyes (David et al., 2007; Abdelmalek et al. 2006). The chemical reactions that degrade such pollutants in wastewater are generally effective in removing color, odor, and carbolic acid, and also decrease biochemical oxygen demand (BOD) and chemical oxygen demand (COD). Such reactions also lead to the destruction and inactivation of viruses, yeast (Yu et al. 2005), and bacteria (Abou-Ghazala et al. 2002; Sato et al. 1996).

Generally, degradation of toxic organic compounds in water is effected primarily by three processes: high-energy electrons, the action of ozone, and UV light. High-energy electrons derived from NTPs provide sufficient energy to dissociate contaminant molecules and produce free radicals, e.g., H_2O dissociating into ˙OH and ˙H. Bubbling oxygen gas through the reactor at the point of discharge results in production of ˙O by dissociation of O_2 and also boosts the rate of production of ˙OH and ˙H. The interaction among these free radicals initiates chain reactions with each other or with water contaminants (and free radicals of contaminants) and results in production of O_3 and H_2O_2. The production efficiency of ˙OH in water is relatively high when ozone and UV light are present. As a result, such chemically reactive species, i.e., those with high-energy electrons (˙H, ˙O, O_3, H_2O_2, and particularly ˙OH) attack and degrade toxic organic compounds in water. The importance of these mechanisms is strongly dependent on the intensity of the energy input to the system being treated. To enhance the degradation rate and the removal efficiency of organic pollutants, many researchers introduce catalysts such as Fe^{2+} ion into NTPs systems. In so doing, they utilize the Fenton reaction, wherein, hydrogen peroxide produced by the discharge plasma reacts with ferrous ion to produce other hydroxyl radicals (Grymonpré et al. 2001). In addition, heterogeneous adsorbants, such as activated carbon (Grymonpré et al. 2003), alumina, and silica gel (Malik 2003), have also been used to promote the degradation of organic contaminants via adsorption. However, such heterogeneous adsorbants, if present in high concentrations, only result in poor degradation efficiency of organic pollutants. Some researchers (Wang et al. 2008; Lukes et al. 2005; Hao et al. 2007) have introduced improved processes that rely on photocatalytic TiO_2 particles. When introduced into NTPs processes, TiO_2 particles produce more chemically active species, especially ˙OH radicals, which accelerate the degradation of organic compounds. However, photocatalytic techniques require prior catalyst–contaminant separation. Recently, magnetite (Fe_3O_4) has attracted scientific interest as a means to purify wastewater (Lei et al. 2007); it retains the benefit of being easily separated and large aggregates are avoided (Moura et al. 2005).

The mechanisms by which toxic organic molecules are degraded are as follows:

$$H_2O + e \rightarrow \text{˙OH} + \text{˙H} \qquad (28)$$

$$O_2 + e \rightarrow \text{˙O} + \text{˙O} \qquad (29)$$

$$\text{˙O} + O_2 \rightarrow O_3 \qquad (30)$$

$$\text{˙OH} + H_2O \rightarrow 2\text{˙OH} \qquad (31)$$

$$\cdot OH + \cdot O \rightarrow H_2O_2 \qquad (32)$$

$$H_2O_2 \rightarrow \cdot OH + \cdot OH \qquad (33)$$

$$O_3 + H_2O_2 \rightarrow \cdot OH + \cdot HO_2 + O_2 \qquad (34)$$

$$\cdot OH + Pollutants \rightarrow Products \qquad (35)$$

To date, applications of NTPs chemistry for wastewater treatment have mainly been limited to laboratory experiments. Future work is needed to develop the comprehensive chemical theory for how NTPs act in water.

3.3 NTPs Chemistry for Quality of Life and Safety

Evidently, all organic compounds can be decomposed in a thermal plasma at high temperatures. In many cases, employing various active species generated in NTPs chemistry may be the preferred method to more economically destroy a host of pollutant problems: toxic substance, dioxins, solvent vapors, pesticides involved in food safety, disinfection for biomaterials, and chemical or biological warfare, such as nerve gases, choking agents, sulfur mustard, etc.

3.3.1 Sterilization

Sterilization is critical in food processing, medical equipment production and handling, packaging of medicine, etc. Results of recent work prove that NTPs are also effective for sterilizing surfaces sufficient to satisfy various sanitary requirements that involve: food safety (Susanne et al. 2007), biomaterials and medical equipment production, and handling and packaging systems (Martinez et al. 2007; Masayuki et al. 2007; Montie et al. 2000; Pérez-Martinez et al., 2007). NTPs chemistry is not only capable of killing bacteria and viruses such as *E. coli bacillus* (Lee et al. 2005; Liu et al. 2008), *Staphylococcus aureus* (Chau et al. 1996), *Bacillus subtilis* (Choi et al. 2006), *Aspergillus niger* spores (Trompeter et al. 2002; Kiel et al. 2002), and yeast (Yu et al. 2005), but also capable of removing dead bacteria and viruses from the surface of objects being sterilized for food safety. There are several mechanisms that may be responsible for the sterilization: interaction of UV light with the DNA of the cells, removal of cellular material (i.e., etching) by reactive species (oxygen atoms and ·OH), and the interaction of these two mechanisms. These mechanisms result in disruption of cell membranes and rupture of DNA, sometimes to the point of cell lysis and erosion of the microorganism (Lee et al. 2005; Cvelbar et al. 2006; Chau et al. 1996; Laroussi et al. 2002), followed by deactivation of the bacteria/microbes.

In addition to use in sterilization, NTPs have also been reported to be effective in removing or cleaning pesticide residues (Clothiaux et al. 1984; Guo et al. 2007;

Herrmann et al. 1999), and effecting insect control (Roe et al. 2003) in food safety applications, without noticeable loss of Vitamin C (Hodgins et al. 2002). Recently, our preliminary research on degradation of pesticides by glow discharge plasmas and DBD also indicate that omethoate and dichlorvos are degraded >80% within 2 min of exposure in these systems. Degradation of the pesticides occurs through a free radical mechanism; intermediates and final products were analyzed using gas chromatography-mass spectrometry (GC-MS; Guo et al. 2007).

3.3.2 Decomposition of CBW agents

Recent increases in terrorist activity and the threat of potential use of agents of chemical and biological warfare (CBW) have led to demand for rapid and reliable methods for degrading such agents. One potential answer to this challenging problem is atmospheric pressure plasma jet (APPJ). APPJ is a non-thermal, glow discharge plasma operating at atmospheric pressure capable of producing reactive species, and propelling them onto contaminated surfaces (Laroussi 2002; Herrmann et al. 2002). APPJ is in development for decontamination of CBW agents by several groups; targets include the following types of CBW agents: mustard blister chemicals, VX nerve toxicants, and anthrax spores. Herrmann et al. (1999, 2002) investigated the decontamination of 2-chloroethyl phenyl sulfide (to simulate sulfur mustard), malathion (a commercially available organophosphorous insecticide used to simulate VX nerve agents), and *Bacillus globigii* (BG; to simulate anthrax spores). Results indicate that a kill efficiency for BG spores is remarkable; the ratio of the number of remaining viable spores to the initial one exceeds 10^{-7} within 30 sec at an effluent temperature of 175°C. The kill efficiency is attained ten times faster than the dry heat administered at the same temperature. This treatment process produced primarily the oxidation product malaoxon, from malathion. Moeller et al. (2000) investigated the destruction of dimethyl-methyl phosphonate (DMMP) using NTPs and conducted off-gas monitoring during the procedure. Results showed that the decontamination treatment did much more than only evaporate DMMP; the treatment resulted in chemical modification of the contaminant. The products that were off-gassed and trapped were primarily phosphorus-containing compounds; much of the carbon content of the contaminant appeared to be oxidized into harmless polymeric and other nonvolatile forms (such as carboxylic acids). Hong et al. (2005) investigated decomposition of phosgene ($COCl_2$) by microwave plasma-torch generated at atmospheric pressure. Their results indicated that the efficiency of phosgene's destruction is nearly 100%; conversion products included Cl_2, CO, and CO_2, and Cl_2 was completely converted to HCl by addition of water. Other researchers have investigated the decomposition of CBW agents by DBD (Laroussi et al. 2000, 2002; Clothiaux et al. 1984).

Thus, NTPs chemistry has the potential to compete with, or replace, traditional methods that are now used for microbial sterilization or pesticide removal for food safety purposes, surface modification of biomaterials, and degradation of CBW agents for safety. In future work, more theoretical studies are needed to better define the chemical reaction mechanisms responsible for NTPs degradation of targeted pollutants.

3.4 NTPs Chemistry for Solid Waste Disposal

Solid waste from industry, agriculture, medical facilities, and waste that results from everyday human activities is the focus of this section. If not properly handled, solid waste (especially medical waste) may do great harm to human health and the environment. Generally, solid waste is treated by incineration in a furnace that is oxygen deficient. Unfortunately, this method may secondarily introduce highly toxic substances such as dioxins, furans, NO_x, and CO, among others.

There are a few literature reports that show improvements in solid waste treatment, particularly use of NTPs chemistry for toxic removal of solid substances; however, these studies are limited to laboratory experiments (Zhou et al. 2001). Zhou et al. (2003) experimented with NTPs for toxic removal of dioxin-containing fly ash. Results showed that the destruction efficiency for dioxin congeners ranges from 20 to 80%. Mizuno (2000) reported that the removal efficiency for some dioxins was >90% when incinerating plastic lumber and vinyl waste for a 15-min period. Rutberg et al. (2002) describes technology designed for use in disinfecting hazardous medical waste using a plasma chemical process, which employs high-temperature mineralization. This method used a low-temperature plasma, produced in a plasma generator, as an additional source of heat energy. The experiment indicated that this method not only neutralized medical waste, but also reduced the volume of waste being buried by 50–400 times.

In short, NTPs can be applied to destroy secondary toxic substances including dioxins, furans, NO_x, and CO after burning solid waste. When NTPs are used to destroy solid waste, the process is improved if the concentration and the volume of waste are reduced by using a burner before NTPs treatment. Moreover, the destruction efficiency of secondary toxic substances is improved if the NTPs is generated using a pulsed corona discharge method.

4 Future Trends

NTPs chemistry has been utilized in several ways and with several different media to achieve improved air quality (indoor air, VOCs, and diesel exhaust), water purity, and food safety standards (sterilization, pesticide degradation, and insect control). NTPs chemistry is effective, and also has these following additional attributes: high efficiency, low-energy consumption, and its utilization do not produce secondary pollutants. Use of certain catalysts in combination with NTPs chemistry improves the selectivity and the energy efficiency of treatment processes. Exactly how these catalysts work from a mechanistic standpoint, however, is unclear. A better understanding is needed on how NTPs work to destroy or degrade pollutant molecules. Once this is known it is expected that various new applications involving NTPs chemistry will surface.

It is recommended that future investigations of NTPs chemistry in the field of environmental pollutants abatement should (1) improve the understanding of the interaction of NTPs and parent molecules that are abatement targets, (2) clarify the chemical mechanisms by which NTPs degrade pollutants, and (3) accelerate commercialization of NTPs equipment to assist in the green revolution of the future – energy efficiency, and without causing secondary pollution.

5 Summary

Over the past several decades, interest in environmental pollutants abatement has greatly increased. This interest is derived from growing concern about environmental pollution and the serious deterioration of many ecosystems as a result of environmental pollution. NTPs chemistry is a proven and effective tool both for decomposing a range of pollutants and for cleaning contaminated surfaces. NTPs chemistry has crucial advantages such as high energy efficiency, higher treatment effectiveness, effective treatment in more confined spaces, and near zero-emissions. When NTPs chemistry is combined with the use of certain catalysts, synergetic pollution abatement results may be achieved; however, the mechanism by which the synergy occurs is still unclear. The purpose of this paper is to provide a basic understanding of NTPs chemistry, including the commonly employed chemical mechanisms, examples of NTPs chemistry applications, and an opinion on the future for NTPs chemistry in the field of environmental pollutants abatement.

Acknowledgments The authors are grateful to Professor L. Zhou for her valuable discussions and comments. We are also grateful to David M. Whitacre for his suggestions during editing. This work was supported by the National Natural Science Foundation of China, No. 30571636.

References

Abdelmalek F, Ghezzar MR, Belhadj M, Addou A., Brisset JL (2006) Bleaching and degradation of textile dyes by nonthermal plasma process at atmospheric pressure. Ind Eng Chem Res 45(1):23–29.

Abou-Ghazala A, Katsuki S, Schoenbach KH, Dobbs FC, Moreira KR (2002) Bacterial decontamination of water by means of pulsed-corona discharges. IEEE Trans Plasma Sci 30(4):1449–1453.

Alexandrov SE, Hitchman ML (2005) Chemical vapor deposition enhanced by atmospheric pressure non-thermal non-equilibrium plasmas. Chem Vap Deposition 11(11–12):457–468.

Anna EW, Whitehead JC, Zhang K (2007) The removal of dichloromethane from atmospheric pressure air streams using plasma-assisted catalysis. Appl Catal B Environ 72(3–4):282–288.

Chang R (2002) Chemistry, 7th Ed. McGraw-Hill, New York, NY, p. 356.

Chang JS (2003) Next generation integrated electrostatic gas cleaning systems. J Electrostat 57(3–4):273–291.
Chang MB, Lee CC (1995) Destruction of formaldehyde with dielectric barrier discharge plasma. Environ Sci Technol 29(1):181–186.
Chang MB, Tseng TD (1996) Gas-phase removal of H_2S and NH_3 with dielectric barrier discharges. J Environ Eng 122(1):41–46.
Chau TT, Kwan CK, Gregory B, Francisco M (1996) Microwave plasmas for low-temperature dry sterilization. Biomaterial 17(13):1273–1277.
Choi JH, Han I, Baik HK, Lee MH, Han DW, Park JC, Lee IS, Song KM, Lim YS (2006) Analysis of sterilization effect by pulsed dielectric barrier discharge. J Electrostat 64(1):17–22.
Chu P K, Chen JY, Wang LP, Huang N (2002) Plasma-surface modification of biomaterials. Mater Sci Eng R: Rep 36(5–6):143–206.
Clements JS, Mizuno A, Finney WC, Davis RH (1989) Combined removal of SO_2, NO_x, and fly ash from simulated flue gas using pulsed streamer corona, IEEE Trans Ind Appl 25(1):62–69.
Clothiaux EJ, Koropchak JA, Moore RR (1984) Decomposition of an organophosphorus material in a silent electrical discharge. Plasma Chem Plasma Process 4(1):15–20.
Cvelbar U, Vujošević D., Vratnica Z, Mozetič M (2006) The influence of substrate material on bacteria sterilization in an oxygen plasma glow discharge. J Physiol D Appl Physiol 39(16):3487–3493.
David M Avaly D, Georges KY, Brisset JL (2007) Postdischarge long life reactive intermediates involved in the plasma chemical degradation of an azoic dye. IEEE Trans Plasma Sci. 35(2):444–453.
Demidiouk V, Chae JO (2005) Decomposition of volatile organic compounds in plasma-catalytic system. IEEE Trans Plasma Sci 33(1):157–161.
Deng SB, Le ZP, Ruan R (2007) Nonthermal plasma synthesis of ammonia using renewable hydrogen. Abstracts of Papers, 234th ACS National Meeting, Boston, MA, United States, August 19–23.
Dyuzhev GA, Ioffe AF (2002) Low-temperature plasma and fullerenes. Plasma Devices Oper 10(2):63–98.
Eichwald O, Yousfi M, Hennad A, Benabdessadok MD (1997) Coupling of chemical kinetics, gas dynamics, and charged particle kinetics models for the analysis of NO reduction from flue gases. J Appl Phys 82(10):4781–4794.
Eliasson B, Kogelschatz U (1991a) Nonequilibrium volume plasma chemical processing. IEEE Trans Plasma Sci 19(6):1063–1077.
Eliasson B, Kogelschatz U (1991b) Modeling and applications of silent discharge plasmas. IEEE Trans Plasma Sci 19(2):309–323.
Fateev A, Leipold F, Kusano Y, Stenum B, Tsakadze E, Bindslev H (2005) Plasma Chemistry in an atmospheric pressure Ar/NH_3 dielectric barrier discharge. Plasma Process Polym 2(3):193–200.
Forst L, Conroy LM (1998) Health effects and exposure assessments of VOCs. In: Rafson HJ (ed) Odor and VOC control handbook. McGraw-Hill, New York, NY, pp. 3.1–3.27.
Futamura S, Sugasawa M (2008) Additive effect on energy efficiency and byproduct distribution in VOC decomposition with nonthermal plasma. IEEE Trans Ind Appl 44(1):40–45.
Futamura S, Yamamoto T (1997) Byproduct identification and mechanism determination in plasma chemical decomposition of trichloroethylene. IEEE Trans Ind Appl 33(2):447–453.
Futamura S, Zhang AH, Yamamoto T (1997) The dependence of nonthermal plasma behavior of VOCs on their chemical structures. J Electrostat 42(1):51–62.
Futamura S, Zhang A, Prieto G, Yamamoto T (1998) Factors and intermediates governing byproduct distribution for decomposition of butane in nonthermal plasma. IEEE Trans Ind Appl 34(5):967–974.
Futamura S, Einaga H, Kabashima H, Hwan LY (2004) Synergistic effect of silent discharge plasma and catalysts on benzene decomposition. Catal Today 89(1):89–95.
Grymonpré DR, Sharma AK, Finney WC, Locke BR (2001) The role of Fenton's reaction in aqueous phase pulsed streamer corona reactors. Chem Eng J 82(1–3):189–207.

Grymonpré DR, Finney WC, Clark RJ, Locke BR (2003) Suspended activated carbon particles and ozone formation in aqueous-phase pulse corona discharge reactors. Ind Eng Chem Res 42(21):5117–5134.

Guo LM, Bai YH, Chen JR (2007) Study on degradation of dichlorvos by nonthermal plasma. Proceedings of the mainland Taiwan Environmental Sustainable Development Academic Conference, Xi'an 10, P.R. China, pp. 171–176.

Hao XL, Zhou MH, Lei LC (2007) Non-thermal plasma-induced photocatalytic degradation of 4-chlorophenol in water. J Hazard Mater 141(3):475–482.

Herrmann HW, Henins I, Park J, Selwyn GS (1999) Decontamination of chemical and biological warfare (CBW) agents using an atmospheric pressure plasma jet (APPJ). Phys Plasmas 6(5):2284–2289.

Herrmann HW, Selwyn GS, Henins I, Park J, Jeffery M, Williams JM (2002) Chemical warfare agent decontamination studies in the plasma decon chamber. IEEE Trans Plasma Sci 30(4):1460–1470.

Higashi M, Uchida S, Suzuki N, Fujii KI (1992) Soot elimination and NO_x and SO_x reduction in diesel-engine exhaust by a combination of discharge plasma and oil dynamics. IEEE Trans Plasma Sci 20(1):1–12.

Hodgins AM, Mittal GS, Griffiths MW (2002) Pasteurization of fresh orange juice using low-energy pulsed electrical field. J Food Sci 67(6):2294–2299.

Hong YC, Uhm HS, Kim HS (2005) Decomposition of phosgene by microwave plasma-torch generated at atmospheric pressure. IEEE Trans Plasma Sci 33(2):958–963.

Hsieh LT, Lee WJ, Chen CY, Wu YPG, Chen SJ, Wang YF (1998) Decomposition of methyl chloride by using an RF plasma reactor. J Hazard Mater 63(1):69–90.

Ingelsten HH, Hildesson A, Fridell E, Skoglundh M (2004) The influence of surface acidity on NO_2 reduction by propane under lean conditions. J Mol Catal A-Chem 209(1–2):199–207.

Jarrige J, Vervisch P (2007) Decomposition of gaseous sulfide compounds in air by pulsed corona discharge. Plasma Chem Plasma Process 27(3):241–255.

Jones AP (1999) Indoor air quality and health. Atmos Environ 33(28):4535–4564.

Joshi AA, Locke BR, Arce P, Finney WC (1995) Formation of hydroxyl radicals, hydrogen peroxide and aqueous electrons by pulsed streamer corona discharge in aqueous solution. J Hazard Mater 41(1):3–30.

Ken Y, Kensuke K, Takashi H, Hiroshi M, Shinji K, Hiroshi M, Toru Y (2007) Efficient decomposition of NO by ammonia radical-injection method using an intermittent dielectric barrier discharge. Thin Solid Films 515(9):4278–4282.

Khacef A, Cormier JM (2006) Pulsed sub-microsecond dielectric barrier discharge treatment of simulated glass manufacturing industry flue gas: Removal of SO_2 and NO_x. J Phys D Appl Phys 39(6):1078–1083.

Khacef A, Cormier JM, Pouvesle JM (2002) NO_x remediation in oxygen-rich exhaust gas using atmospheric pressure non-thermal plasma generated by a pulsed nanosecond dielectric barrier discharge. J Phys D Appl Phys 35(17):1491–1498.

Kiel JL, Sutter RE, Mason PA, Parker JE, Morales PJ, Stribling LJV, Alls JL, Holwitt EA, Seaman RL, Mathur SP (2002) Direct killing of anthrax spores by microwave-induced cavitation. IEEE Trans Plasma Sci 30(4):1482–1488.

Kim HH (2004) Nonthermal plasma processing for air-pollution control: A historical review, current issues, and future prospects. Plasma Process Polym 1(2):91–110.

Kim HH, Takashima K, Katsura S, Mizuno A (2001) Low-temperature NO_x reduction processes using combined systems of pulsed corona discharge and catalysts. J Phys D Appl Phys 34(4):604–613.

Kim HH, Ogata A, Futamura S (2008) Oxygen partial pressure-dependent behavior of various catalysts for the total oxidation of VOCs using cycled system of adsorption and oxygen plasma. Appl Catal B Environ 79(4):356–367.

Kirkpatrick M, Locke BR (2005) Hydrogen, oxygen, and hydrogen peroxide formation in electrohydraulic discharge. Ind Eng Chem Res 44(12):4243–4248.

Kogelschatz U (2003) Dielectric-barrier discharges: Their history, discharge physics, and industrial applications. Plasma Chem Plasma Process 23(1):1–46.

Koutsospyros AD, Shu-Min Y, Christodoulatos C, Becker K (2005) Plasmochemical degradation of volatile organic compounds (VOC) in a capillary discharge plasma reactor. IEEE Trans Plasma Sci 33(1):42–49.

Kuroki T, Takahashi M, Okubo M, Yamamoto T (2002) Single-stage plasma-chemical process for particulates, NO_x, and SO_x simultaneous removal. IEEE Trans Ind Appl 38(5):1204–1209.

Laguardia L, Vassallo E, Cappitelli F, Mesto E, Cremona A, Sorlini C, Bonizzoni G (2005) Investigation of the effects of plasma treatments on biodeteriorated ancient paper. Appl Surf Sci 252(4):1159–1166.

Laroussi M (2002) Nonthermal decontamination of biological media by atmospheric pressure plasmas: Review, analysis, and prospects. IEEE Trans Plasma Sci 30(4):1409–1415.

Laroussi M, Alexeff I, Kang WL (2000) Biological decontamination by nonthermal plasma. IEEE Trans Plasma Sci 28(1):184–188.

Laroussi M, Richardson JP, Dobbs FC (2002) Effects of nonequilibrium atmospheric pressure plasmas on the heterotrophic pathways of bacteria and on their cell morphology. Appl Phys Lett 81(4):772–774.

Lee HW, Chang MB (2001) Gas-phase removal of acetaldehyde via packed-bed dielectric barrier discharge reactor. Plasma Chem Plasma Process 21(3):329–343.

Lee KY, Park B.J., Lee DH, Lee IS, Hyun SO, Chung KH, Park JC (2005) Sterilization of *Escherichia coli* and MRSA using microwave-induced argon plasma at atmospheric pressure. Surf Coat Technol 19(1–3):35–38.

Lei LC, Hao XL, Zhang XW, Zhou MH (2007) Wastewater treatment using a heterogeneous magnetite (Fe_3O_4) non-thermal plasma process. Plasma Process Polym 4(4):455–462.

Li R, Chen JR (2006) Studies on wettability of medical poly (vinyl chloride) by remote argon plasma. Appl Surf Sci 252(14):5076–5082.

Li JH, Ke R, Li W, Hao JM (2007) A comparison study on non-thermal plasma-assisted catalytic reduction of NO by C_3H_6 at low temperatures between Ag/USY and Ag/Al_2O_3 catalysts. Catal Today 126(3–4):272–278.

Liu HX, Chen JR, Yang LQ, Zhou Y (2008) Long-distance oxygen plasma sterilization: Effects and mechanisms. Appl Surf Sci 254(6):1815–1821.

Locke BR, Sato M, Sunka P, Hoffmann MR, Chang JS (2006) Electrohydraulic discharge and nonthermal plasma for water treatment. Ind Eng Chem Res 45(3):882–905.

Louis AR (2005) Nonthermal plasma applications to the environment: Gaseous electronics and power conditioning. IEEE Trans Plasma Sci 33(1):129–137.

Lovejoy ER, Murrells TP, Ravishankara AR, Howard CJ (1990) Oxidation of carbon disulfide by reaction with hydroxyl. 2. Yields of hydroperoxyl and sulfur dioxide in oxygen. J Phys Chem 94(6):2386–2393.

Lukes P, Appleton AT, Locke BR (2004) Hydrogen peroxide and ozone formation in hybrid gas-liquid electrical discharge reactors. IEEE Trans Ind Appl 40(1):60–67.

Lukes P, Clupek M, Sunka P, Peterka F, Sano T, Negishi N, Matsuzawa S, Takeuchi K (2005) Degradation of phenol by underwater pulsed corona discharge in combination with TiO_2 photocatalysis. Res Chem Intermediat 31(4–6):285–294.

Ma H, Chen P, Ruan R (2001) H_2S and NH_3 removal by silent discharge plasma and ozone combo-system. Plasma Chem Plasma Process 21(4):611–624.

Magne L, Pasquiers S (2005) LIF spectroscopy applied to the study of non-thermal plasmas for atmospheric pollutant abatement. C R Physique 6(8):908–917.

Magureanu M, Mandache NB, Parvulescu VI (2007) Improved performance of non-thermal plasma reactor during decomposition of trichloroethylene: Optimization of the reactor geometry and introduction of catalytic electrode. Appl Catal B Environ 74(3–4):270–277.

Malik MA (2003) Synergistic effect of plasmacatalyst and ozone in a pulsed corona discharge reactor on the decomposition of organic pollutants in water. Plasma Sources Sci Technol 12(4):26–32.

Marotta E, Scorrano G, Paradisi C (2005) Ionic reactions of chlorinated volatile organic compounds in air plasma at atmospheric pressure. Plasma Process Polym 2(3):209–217.

Martinez H, Rodriguez-Lazcanol Y, Castillo F (2007) Comparative study on the decomposition process of N-isopropyl-acrylamide in He, N_2 and air plasmas. Plasma Sources Sci Technol 16(3):427–433.

Masayuki A, Toshifumi Y, Takayuki W, Junzou K, Yoshitake S (2007) Application to cleaning of waste plastic surfaces using atmospheric non-thermal plasma jets. Thin Solid Films 515(9):4301–4307.

Masuda S (1988) Pulse corona induced plasmachemical process: A horizon of new plasmachemical technologies. Pure Appl Chem 60(5):727–731.

McQuarrie DA, Rock PA (1984) General chemistry. Freeman, New York, NY, p. 624.

McTaggart FK (1967) Plasma chemistry in electrical discharges. Elsevier, The Netherlands, p. 1.

Mizuno A (2000) Electrostatic precipitation. IEEE Trans Dielectr Electr Insul 7(5):615–624.

Mizuno A (2007) Industrial applications of atmospheric non-thermal plasma in environmental remediation. Plasma Phys Control Fusion 49(5A):A1–A15.

Moeller TM, Alexander ML, Engelhard MH, Gaspar DJ, Luna ML, Irving PM (2000) Surface decontamination of simulated chemical warfare agents using a nonequilibrium plasma with off-gas monitoring. IEEE Trans Plasma Sci 28(4):1454–1459.

Mok YS, Nam CM, Cho MH, Nam IS (2002) Decomposition of volatile organic compounds and nitric oxide by nonthermal plasma discharge processes. IEEE Trans Plasma Sci 30(1):408–416.

Montie TC, Kelly-Wintenberg K, Roth JR (2000) An overview of research using the on atmosphere uniform glow discharge plasma (OAUGDP) for sterilization of surfaces and materials. IEEE Trans Plasma Sci 28(1):41–50.

Moura FCC, Araujo MH, Costa RCC, Fabris JD, Ardisson JD, Macedo WAA, Lago RM (2005) Efficient use of Fe metal as an electron transfer agent in a heterogeneous Fenton system based on Fe^0/Fe_3O_4 composites. Chemosphere 60(8):1118–1123.

Nair SA, Nozaki T, Okazaki K (2007) Methane oxidative conversion pathways in a dielectric barrier discharge reactor – Investigation of gas phase mechanism. Chem Eng J 132(1–3):85–95.

Nicole BS, Pierre T, Aurore R, François J, Stéphane P (2005) Removal of 2-Heptanone by dielectric barrier discharges – The effect of a catalyst support. Plasma Process Polym 2(3):256–262.

Oda T, Kato T, Takahashi T, Shimizu K (1998) Nitric oxide decomposition in air by using nonthermal plasma processing with additives and catalyst. IEEE Trans Ind Appl 34(2):268–272.

Ogata A, Einaga H, Kabashima H, Futamura S, Kushiyama S, Kim HH (2003) Effective combination of nonthermal plasma and catalysts for decomposition of benzene in air. Appl Catal B Environ 46(1):87–95.

Oh SM, Kim HH, Einaga H, Ogata A, Futamura S, Park DW (2006) Zeolite-combined plasma reactor for decomposition of toluene. Thin Solid Films 508(1–2):418–422.

Ono R, Oda T (2000) Measurement of hydroxyl radicals in an atmospheric pressure discharge plasma by using laser-induced fluorescence. IEEE Trans Ind Appl 36(1):82–86.

Penetrante BM, Hsiao MC, Merritt BT, Vogtlin GE, Wallman PH (1995) Comparison of electrical discharge techniques for nonthermal plasma processing of NO in N_2. IEEE Trans Plasma Sci 23(4):679–687.

Penetrante BM, Hsiao MC, Bardsley JN, Merritty BT, Vogtliny GE, Kuthiz A, Burkhartz CP, Baylessz JR (1997) Identification of mechanisms for decomposition of air pollutants by nonthermal plasma processing. Plasma Sources Sci Technol 6(3):251–259.

Pérez-Martinez JA, Pena-Eguiluz R, López-Callejas R, Mercado-Cabrera A, Valencia RA, Barocio SR, Benítez-Read JS, Pacheco-Sotelo JO (2007) An RF microplasma facility development for medical applications. Surf Coat Technol 201(9–11):5684–5687.

Peter AG, Whitehead JC, Wu JH (2007) Adaptive control for NO_x removal in non-thermal plasma processing. Plasma Process Polym 4(5):556–562.

Petrovic ZL, Makabe T (1998) Nonequilibrium plasmas for material processing in microelectronics. Adv Mater Proc 282(2):47–56.

Roe RM, Long SY, Bourham MA, Bures BL, Gray TK (2003) Use of atmospheric plasma for insect control. Proceedings of Beltwide Cotton Conferences. National Cotton Council, Atlanta, GA, pp. 1150–1156.

Rudolph R, Francke KP, Miessner H (2002) Concentration dependence of VOC decomposition by dielectric barrier discharges. Plasma Chem Plasma Process 22(3):401–412.

Rutberg PG, Bratsev AN, Safronov AA, Surov AV, Schegolev VV (2002) The technology and execution of plasma-chemical disinfection of hazardous medical waste. IEEE Trans Plasma Sci 30(4):445–1448.

Sahni M, Finney WC, Locke BR (2005) Degradation of aqueous phase polychlorinated biphenyls (PCB) using pulsed corona discharges. J Adv Oxid Technol 8(1):105–111.

Sano T, Negishi N, Sakai E, Matsuzawa S (2006) Contributions of photocatalytic/catalytic activities of TiO_2 and γ-Al_2O_3 in nonthermal plasma on oxidation of acetaldehyde and CO. J Mol Catal A-Chem 245(1–2):235–241.

Sato M, Ohgiyama T, Clements JS (1996) Formation of chemical species and their effects on microorganisms using a pulsed high-voltage discharge in water. IEEE Trans Ind Appl 32(1):106–112.

Savinov SY, Lee H, Song HK, Na BK (2003) The effect of vibrational excitation of molecules on plasmachemical reactions involving methane and nitrogen. Plasma Chem Plasma Process 23(1):159–173.

Schutze A, Jeong JY, Babayan SE, Park J, Selwyn GS, Hicks RF (1998) The atmospheric-pressure plasma jet: A review and comparison to other plasma sources. IEEE Trans Plasma Sci 26(6):1685–1694.

Shi Y, Ruan JJ, Wang X, Li W, Tan TE (2006) Evaluation of multiple corona reactor modes and the application in odor removal. Plasma Chem Plasma Process 26(2):187–196.

Siemens W (1857) Ueber die elektrostatische Induction und die Verzögerung des Stroms in Flaschendräten. Poggendorffs Ann Phys Chem 102:66–120.

Storch DG, Kushner MJ (1993) Destruction mechanisms for formaldehyde in atmospheric pressure low temperature plasmas. J Appl Phys 73(1):51–56.

Subrahmanyam C, Renken A, Kiwi-Minsker L (2007) Novel catalytic non-thermal plasma reactor for the abatement of VOCs. Chem Eng J 134(1–3):78–83.

Šunka P (2001) Pulse electrical discharges in water and their applications. Phys Plasmas 8(5):2587–2594.

Šunka P, Babicky V, lupek M, Lukeš P, Šimek M, Schmidt J, ernak M (1999) Generation of chemically active species by electrical discharges in water. Plasma Sources Sci Technol 8(2):258–265.

Susanne S, Thorsten A, Stefan T, Sybille N, Dietrich K, Andreas S, Reinhold C (2007) Effects of pulsed electric field treatment of apple mash on juice yield and quality attributes of apple juices. Innovat Food Sci Emerg Technol 8(1):127–134.

Tezuka M, Iwasaki M (2001) Plasma-induced degradation of aniline in aqueous solution. Thin Solid Films 386(2):204–207.

Thevenet F, Guaitella O, Puzenat E (2007) Oxidation of acetylene by photocatalysis coupled with dielectric barrier discharge. Catal Today 122(1–2):186–194.

Tonks L, Langmuir I (1929) Oscillations in ionized gases. Phys Rev 33(2):195–210.

Trompeter FJ, Neff WJ, Franken O, Heise M, Neiger M, Liu SH, Pietsch GJ, Saveljew AB (2002) Reduction of *Bacillus subtilis* and *Aspergillus niger* spores using nonthermal atmospheric gas discharge. IEEE Trans Plasma Sci 30(4):1416–1422.

Van Durme J, Dewulf J, Sysmans W, Leys C, Van Langenhove H (2007) Abatement and degradation pathways of toluene in indoor air by positive corona discharge. Chemosphere 68(10):1821–1829.

Wallis AE, Whitehead J, Zhang C (2007) Plasma-assisted catalysis for the destruction of CFC-12 in atmospheric pressure gas streams using TiO_2. Catal Lett 113(1–2):29–33.

Wang HJ, Li J, Quan X, Wu Y (2008) Enhanced generation of oxidative species and phenol degradation in a discharge plasma system coupled with TiO_2 photocatalysis. Appl Cata B Environ 83(1–2):72–77.

Wen YZ, Jiang XZ (2001) Pulsed corona discharge-induced reactions of acetophenone in water. Plasma Chem Plasma Process 21(3):345–354.

Willberg DM, Lang PS, Hochemer RH, Kratel A, Hoffmann MR (1996) Degradation of 4-chlorophenol, 3,4-dichloroaniline, and 2,4,6-trinitrotoluene in an electrohydraulic discharge reactor. Environ Sci Technol 30(8):2526–2534.

Xu RF, Hu XX, Hu WK, Xu XY (2003) Study of the mechanism of HCHO photocatalyst degraded by nanosized TiO_2. Chem Res Appl 15(5):715–717 (in Chinese).

Yamamoto T (1997) VOC decomposition by nonthermal plasma processing—A new approach. J Electrostat 42(1–2):227–238.

Yamamoto T, Rajanikanth BS, Masaaki O (2003) Performance evaluation of nonthermal plasma reactors for NO oxidation in diesel engine exhaust gas treatment. IEEE Trans Ind Appl 39(6):1608–1613.

Yan K, Kanazawa S, Ohkubo T, Nomoto Y (1999) Oxidation and reduction processes during NO_x removal with corona-induced nonthermal plasma. Plasma Chem Plasma Process 19(3):421–443.

Yan K, van Heesch EJM, Pemen AJM, Huijbrechts PAHJ (2001) From chemical kinetics to streamer corona reactor and voltage pulse generator. Plasma Chem Plasma Process 21(1):107–137.

Yen A, Kim SS, Hecht MH, Frant MS, Murray B (2000) Evidence that the reactivity of the Martian soil is due to superoxide ions. Science 289(5486):1909–1912.

Yoshihiko I, Matsuei U, Hirohumi S (2006) NO_x reduction behavior on alumina with discharging nonthermal plasma in simulated oxidizing exhaust gas. J Chem Technol Biotechnol 81(4):544–552.

Yu H, Xiu ZL, Ren CS, Zhang JL, Wang DZ, Wang YN, Ma TC (2005) Inactivation of yeast by dielectric barrier discharge (DBD) plasma in helium at atmospheric pressure. IEEE Trans Plasma Sci 33(4):1405–1409.

Zhang A, Futamura S, Yamamoto T (1999) Nonthermal plasma chemical processing of bromomethane. J Air Waste Manage Assoc 49(12):1442–1448.

Zhang JB, Zheng Z, Zhang YN, Feng JW, Li JH (2008) Low-temperature plasma-induced degradation of aqueous 2,4-dinitrophenol. J Hazard Mater 154(1–3):506–512.

Zhou YX, Nifuku M, Hajois G, Asada S, Katoh H (2001) An investigation on pulse discharge effect on the surface chemical transformation of fly ash. Thin Solid Films 386(2):195–199.

Zhou YX, Yan P, Cheng ZX, Nifuku M, Liang XD, Guan ZC (2003) Application of non-thermal plasmas on toxic removal of dioxin-contained fly ash. Powder Technol 135–136:345–353.

Environmental Fate and Global Distribution of Polychlorinated Biphenyls

Angelika Beyer and Marek Biziuk

Contents

1. Introduction ... 137
2. Global Distribution ... 139
3. Environmental Fate ... 141
   - 3.1 Biodegradation and Transformation ... 142
   - 3.2 Volatilization ... 146
   - 3.3 Adsorption to Organic Matter ... 147
   - 3.4 Bioaccumulation, Bioconcentration, and Biomagnification ... 147
4. Summary ... 153
   References ... 154

1 Introduction

Polychlorinated biphenyls (PCBs) are synthetic chlorinated organic chemicals. They are environmentally widespread and persistent and are routinely found in air, water, sediments, and soils. Moreover, they accumulate through the food chain from aquatic organisms to fish and to humans. PCBs are complex mixtures of individual chlorobiphenyls. The low reactivity and high chemical stability of the PCBs have made them useful for numerous industrial applications; the qualities that make many individual chlorobiphenyls industrially useful, however, render them more persistent and less environmentally desirable than many other organic chemicals.

There is growing concern about the trace quantities of *highly chlorinated organic compounds* (e.g., dioxins, *PCBs*, and certain pesticides) that exist in diverse environmental media (air, soil, water, and biota). Such halogenated organic compounds enter the food chain from environmental media, mainly through intake of animal or fish fats (meat, fish, and milk), and reach humans and wildlife.

A. Beyer (✉)
Department of Analytical Chemistry, Chemical Faculty, Gdansk University of Technology,
G. Narutowicza Street 11/12, 80-233 Gdansk, Poland
e-mail: angelika.beyer@gmail.com

A common ingredient of these compounds is chlorine, one of the most abundant naturally occurring chemical elements on this planet. Since the discovery of chlorine, the chemical industry has developed many processes and products that either contain chlorine or are produced with its help. However, as the volume and diversity of chlorinated substances has increased, environmental concerns for them has also increased; many chlorinated moieties that are persistent, toxic, and bioaccumulative have been banned (Lopez 2003; Lohmann et al. 2007). Consequently, there is growing scientific, regulatory, and social interest in measuring the levels of chlorinated chemicals in environmental media, and in determining the environmental effects of such contamination.

PCBs are produced by chlorination of biphenyls and comprise a class of 209 individual discrete chemicals, each with 12 carbon atoms. The PCB structure is presented in Fig. 1. Chlorine is substituted for hydrogen atoms in any of the ten numbered positions of this structure.

Other than commercial production, PCBs are also created as thermodynamically stable by-products of certain combustion processes such as incineration of PCB-containing wastes. EPA regulations concerning the storage and disposal of PCBs have specified incineration as the only acceptable method of PCB disposal unless, by reason of the inability to dispose of the waste or contaminated materials in this manner, clearance is obtained from EPA to dispose of materials in another way (EPA 1978). However, the general acceptance of incineration as a means of disposal for PCB-contaminated materials has declined because of concerns about incomplete incineration and the possible formation of highly toxic by-products, such as hydrogen chloride, polychlorinated dibenzodioxins, and polychlorinated dibenzofurans, if the combustion temperature is not held sufficiently high (Arbon et al. 1994; Chuang et al. 1995).

Depending on the degree of PCB chlorination and their physicochemical properties (Fig. 2), PCBs have successfully been used as stable fluid insulators in high-voltage electric transformers, in high-capacity condensers, as heat exchangers, pesticide extenders, adhesives, dedusting agents, components of cutting oils, flame retardants, hydraulic lubricants, and components of plasticizers in paints, inks, toners, and printing inks (Erickson 1997).

As PCBs move through specific environmental compartments, or move from one environmental medium to another, the relative concentrations of individual chlorobiphenyls change. These changes result from physical and chemical processes and selective bioaccumulation and biotransformation of PCBs by living organisms. These processes produce mixtures that are substantially different from the ones

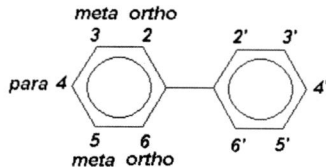

Fig. 1 Structure of polychlorinated biphenyls (PCBs)

Physico-chemical properties of PCBs	• good solubility in nonpolar solvents, oils and fats (high lipophilicity), • low vapor pressure, • non-explosive, • low electrical conductivity, • very high thermal conductivity, • high ignition temperature, • extremely high thermal and chemical resistance (very high stability), • oily liquids whose color darkens and viscosity increases with rising chlorine content.

Fig. 2 Physicochemical properties of PCBs

originally deposited into the environment. The primary reason that aged environmental residues of PCB mixtures differ from the original forms deposited in the environment is that some congeners are more easily degraded than others. In particular, *ortho*-substituted congeners are readily degradable; therefore, they are not as abundant in aged environmental samples. In contrast, the so-called "dioxin-like" PCBs, namely the coplanar (= *non-ortho* substituted) and *mono-ortho*-substituted congeners, are very stable and resist biodegradation and metabolism; these congeners are present in commercial mixtures and in aged environmental samples as well. Therefore, identification and quantification of the PCB mixtures on which the environment has acted is complicated, as is the task of assessing the risk posed by these residual mixtures in the environment.

In this paper, we review the global distribution and environmental fate of PCBs. We emphasize aspects pertaining particularly to bioaccumulation, bioconcentration, and biomagnification of these environmental contaminants. Our review supports the conclusion that the presence of PCBs in the environment may pose long-term public health and ecosystem risks.

2 Global Distribution

Human activities constitute the only source of PCBs in the environment. Hence, sites that are highly contaminated by PCBs tend to exist in industrialized areas. PCBs are ubiquitous in the general environment (Fig. 3) and consequently are transported by wind and water. PCBs bind strongly to organic particles in the water column, atmospheric particulates, sediments, and soil. The deposition of particle-bound PCBs from the atmosphere and sedimentation from water are largely responsible for their accumulation in sediments and soils.

The principal transport route for PCBs through aquatic systems is from waste streams into receiving waters. Currently, the major source of PCB release to surface water is the environmental cycling process (Mackay 1989; Swackhamer and Armstrong 1986). Small amounts of PCBs may enter receiving waters by runoff from accidental spillage of PCB-containing hydraulic fluids, disposal of waste oils

- accidental spills of PCB-containing hydraulic fluids
- improper disposal
- combined sewer overflows, or storm water runoff
- runoff and leaching from PCB-contaminated sewage sludge applied to farmland

- accidental leaks and spills
- release from contaminated soils in landfills and hazardous waste sites
- deposition of vehicular emissions near roadway soil
- land application of sewage sludges containing PCBs

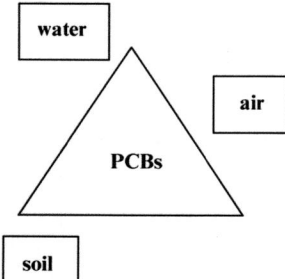

- volatilization from soil and water
- escape from uncontrolled landfills and hazardous waste sites
- incineration of PCB-containing wastes
- leakage from older electrical equipment
- improper waste disposal or spills
- leakage from supposedly closed systems
- incineration of waste
- industrial discharges
- sewage effluents

Fig. 3 How PCBs enter the general environment

into street drains, or from farmland to which sewage sludge containing small quantities of PCBs has been applied (Gan and Berthouex 1994; Gunkel et al. 1995). PCBs are persistent in waters where an equilibrium is established as PCBs are adsorbed from water onto particles or sediments or desorbed back into water or biota. The marine environment is one of the final sinks for PCBs. PCBs do not remain long dissolved in water. Therefore, they volatilize into the atmosphere, and large quantities volatize from lakes and seas, even as PCBs from global air masses are deposited into them (U.S. Department of Health and Human Services et al. 2000; Wolska et al. 2003).

The importance of volatilization to atmospheric concentrations of PCBs is well established. This conclusion is also supported by the estimated Henry's law constants (HLCs) for individual PCB congeners (Thomas 1982). Henry's law constant is a fundamental property that describes the tendency for water to reach an equilibrium between the atmosphere and surface waters. The lack of accurate HLC values is a major problem in modeling the transport and fate of PCBs in the environment, a necessary step for performing risk assessment and developing environmental remediation strategies (Phillips et al. 2008). HLCs have been measured for selected congeners, commonly at 25°C. The largest experimental data sets at this temperature are those of Brunner et al. (1990), who measured HLCs for 58 congeners, and Bamford et al. (2000), who measured HLCs for 26 congeners. However, HLC is a highly temperature-dependent property, and transport and fate modeling requires knowledge of HLCs over a range of environmentally relevant temperature (Phillips et al. 2008).

The toxicity to animal life mainly results from the presence of PCBs in water. The PCBs that exist in water or sediments are filtered through fish, crustaceans, or mollusks via their gills; because these xenobiotics are lipophilic, they make their way into the fatty tissues of these animals. These aquatic animals are, in turn, eaten by birds and mammals, which also accumulate PCBs in their fatty tissues. Such intake at successive trophic levels further concentrates the PCB residues. Such concentration from one trophic level to another is called biomagnification and is further explained later. Yolks rich in fatty material, from eggs laid by bird species near the top of the food chain, often have the highest concentrations of PCBs.

Several approaches are used to study the fate of PCBs in the environment. The main purpose of such studies is to understand contaminant concentrations and fluxes among environmental compartments under a variety of environmental scenarios. Traditional monitoring tools, such as measurements of chemical concentrations, are employed to verify the level of contaminants in a given environmental compartment (air, water, soil, sediment, biota; Borga and Di Guardo 2005). Since the 1970s, microecosystems (microcosms) have also been widely used to gain an understanding of the effects of toxic substances at the ecosystem level; microecosystems, based on this small prototype concept, were constructed for the study of pesticides (Kersting 1975, 1984, 1991; Kersting and Van Wijngaarden 1992). In these studies ideas about quantifying changes that occur in complex systems have evolved (Kersting 1984, 1991). Another approach is to mathematically model the contaminant distribution in the environment by using multimedia models. With this approach, an attempt is made to relate physicochemical properties of a chemical to an understanding of its probable partition and transport in the environment. Models can be purely empirical or mechanistic (Boese et al. 1995; Van Bavel et al. 1996). Borga and Di Guardo (2005), in their paper, illustrate and discuss some of the difficulties encountered when different approaches are used to study environmental distribution of contaminants. Breivik et al. (2007) present an update of a dynamic mass balance approach aimed at presenting the "big picture" of the global historical atmospheric emissions of selected PCBs.

3 Environmental Fate

As a result of their chemical stability, PCBs are environmentally persistent and are among the most prominent and widespread environmental contaminants. When these undesirable features are combined with the low water solubility of PCBs, great concern emerges, because these substances may be (and are) accumulated through the food chain and reach aquatic organisms, fish, and humans. However, it has been discovered, with great interest, that PCBs can be dehalogenated in freshwater and in estuarine sediments (Abramowicz 1994; Bedard et al. 1996; Berkaw et al. 1996; Brown et al. 1987; Brown and Wagner 1990; Fish and Principe 1994; Quensen et al. 1990). The deep interest in this discovery derives from the fact that dehalogenation is expected to detoxify these dangerous substances (Bedard and Van Dort 1997). Another important pathway for PCB loss is volatilization, and colloidal adsorption of PCBs to dissolved organic matter (DOM), which reduces bioavailability (Fig. 4).

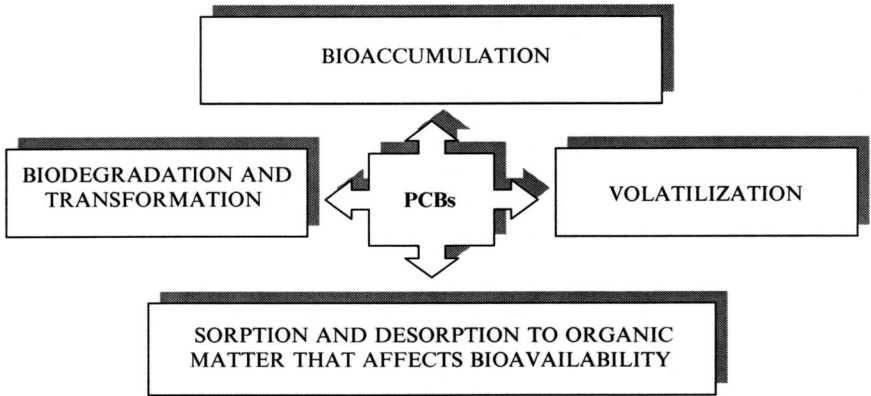

Fig. 4 Processes that affect the environmental fate of PCBs

3.1 Biodegradation and Transformation

Biodegradation occurs primarily through the action of bacteria or other microorganisms, either aerobically or anaerobically. It is the only process known to degrade PCBs in soil or in aquatic environments. The specific processes that achieve degradation are generally either aerobic oxidative dechlorination, or hydrolytic dehalogenation and anaerobic reductive dechlorination. Theoretically, the biological degradation of PCBs should produce CO_2, chlorine, and water. This process involves the removal of chlorine from the biphenyl ring followed by ring cleavage and oxidation of the resulting compound (Boyle et al. 1992).

Different microbes have different capabilities to degrade PCBs. Microbial degradation processes comprise a set of dechlorination reactions that determine which PCB congeners (of the 209 congeners that are theoretically possible) are suitable substrates, which chlorines will be removed from those congeners, and the order in which these chlorines will be removed (Bedard and Quensen 1995). Results of several studies have demonstrated that different PCB dechlorination processes appear to be responsible for microbial dechlorination (Bedard and May 1996; Bedard and Quensen 1995; Brown and Wagner 1990; Sokol et al. 1994), but maximal chlorine removal requires the complementary action of two or more such dechlorination processes (Bedard and Quensen 1995; Quensen et al. 1990). It has been proposed that discrete dechlorinating microorganisms that harbor dehalogenases, with different regiospecificities, are responsible for the various dechlorination processes that have been described (Bedard et al. 1993, 1996, 1997; Bedard and Quensen 1995; Brown and Wagner 1990; Quensen et al. 1990).

For example, aerobic bacteria of the genus *Pseudomonas* degrade PCBs by aerobic oxidative dehalogenation. This process involves the addition of oxygen to the biphenyl ring. The metabolic pathway used by *Micrococcus* sp. resembles that described for *Pseudomonas* sp. and is illustrated in Fig. 5 (Benvinakatti and Ninnekar 1992).

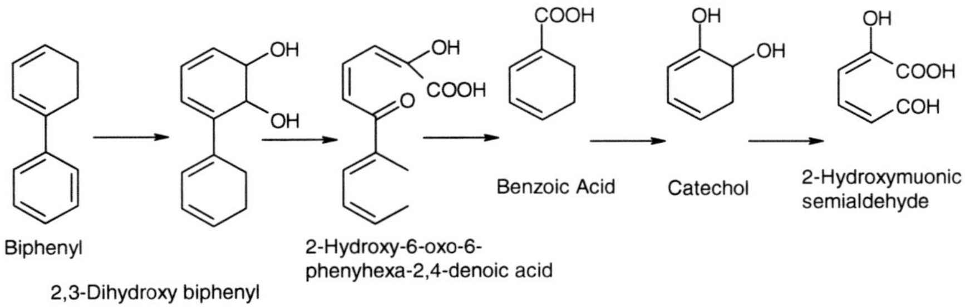

Fig. 5 A possible pathway for the aerobic oxidative dehalogenation of PCBs

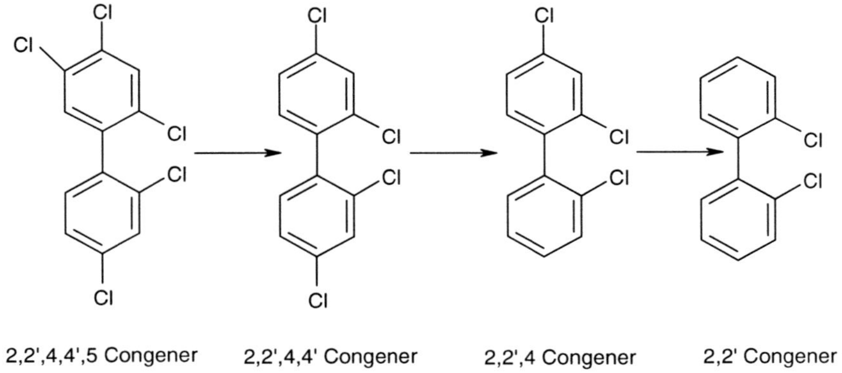

Fig. 6 A potential pathway for anaerobic degradation of highly chlorinated PCB congeners to less chlorinated ones

PCB congeners can be dechlorinated selectively by either aerobic or anaerobic bacteria. However, only the lower chlorinated congeners are amenable to aerobic degradation. Aerobic PCB biodegradation is a cometabolic process, in which another substrate such as biphenyl is required as a carbon and energy source. Alternatively, anaerobic reductive dechlorination occurs only for the most heavily chlorinated PCB congeners. This process may be of selective advantage to microorganisms because it produces a gain in energy for the organism (Jafvert and Rogers 1990).

The rate of anaerobic dechlorination is constrained mainly by rates of chlorine removal from *meta* and *para* positions. Some methanogenic bacteria, for example, preferentially dechlorinate PCBs at the *meta* and *para* positions, with a resulting enhancement of mono-, di-, and tri-chlorinated *ortho*-substituted PCBs (Alder et al. 1993). In anaerobic environments, highly chlorinated biphenyls can undergo reductive dechlorination to less chlorinated congeners. Specifically, monochlorobiphenyls and *ortho*-substituted dichlorobiphenyls degrade in this manner. This type of degradation involves the replacement of chlorine with a hydrogen atom on the biphenyl ring and is illustrated in Fig. 6 (Fish and Principe 1994). Byproducts are less toxic than the parent molecules and can be degraded by aerobic microbes.

Under anaerobic conditions, not only are PCBs with *para*- and *meta*-substituted rings more easily degraded than the *ortho*-substituted compounds, but PCBs containing all chlorines on one ring are biodegraded faster than those that contain chlorines throughout both rings (Abramowicz 1995).

The study of bacterial PCB biodegradation has been considerably assisted by use of tritium-labeled PCBs. High specific activity permits tracing all tritium-labeled PCB biodegradation products. Kim et al. (2004) found that four *Bacillus* sp. strains retained the ability to accumulate and metabolically destroy PCBs.

Different microbial populations have demonstrated differing congener specificity when dechlorinating PCBs or breaking their carbon–carbons bonds (Quensen et al. 1990). However, such bacterial populations appear to display similar degradation patterns. In Woods Pond, at least seven distinct microbial dechlorination activities have been identified and described (Wu et al. 1997a; Table 1). It was shown in this study that individual reactions in sediments exhibited strong temperature dependencies. Moreover, temperature influenced the timing and preferred sequence of dechlorination. Results indicated that several distinct PCB-dechlorinating microbial populations, with different temperature ranges and different dechlorination specificities, exist in these sediments, but no pure cultures have yet been isolated. Van Dort and Bedard (1991) reported the first experimental demonstration of biologically mediated *ortho*-dechlorination of a PCB, and the first stoichiometric conversion of a PCB congener to less-chlorinated forms. However, this process does not appear to be important under natural conditions. Bedard et al. (1997) stimulated a new *para*-dechlorination activity (Process LP) that caused further conversion of Process N products to tri- and tetra-chlorobiphenyls.

To date, there are at least eight distinct, documented reductive dechlorination pathways or processes, each resulting in a different congener distribution profile. These dechlorination processes have been identified from careful examination of congener loss and product accumulation patterns in different sediment samples (Bedard 2003; Wiegel and Wu 2000) and marked using letters: N, LP, M, Q, H, H', T, P. Apart from patterns shown in Table 1, H, H', M, and Q processes have been described (Bedard 2003; Wiegel and Wu 2000). The documented patterns include *meta*- and *para*-dechlorination and also reflect the relatively infrequent observation of *ortho*-dechlorination, in many of the initial sediment studies.

Anaerobic transformation of highly chlorinated congeners into lower ones, in a variable environment, renders them subject to later aerobic microbial degradation, which can oxidatively mineralize lower Cl congeners such as the homologs Cl_1 and Cl_2 to carbon dioxide and water (Bedard et al. 1987; Bedard and Quensen 1995).

A detailed description of the potential for microorganisms to transform PCBs can be found in numerous review articles (Field and Sierra-Alvarez 2008).

Fish are generally considered to have poor capability to biotransform PCBs (Boon et al. 1989; Matthews and Dedrick 1984). However, results of some work suggest that fish do have some capacity to biotransform PCBs (Wong et al. 2002, 2004), and the formation of hydroxylated PCBs (OH-PCBs) in fish has been reported (White et al. 1997). Hydroxylated PCBs were found in a number of fish species from the Great Lakes (Campbell et al. 2003; Li et al. 2003), which presents

Table 1 The microbial reductive dechlorination of 2,3,4,6-tetrachlorobiphenyl in PCB-contaminated sediments from Woods Pond (Lenox, MA)

Process name	Dechlorination reaction	Specificity of dechlorination	Temperature range (°C)	Temperature at which reaction was dominant (°C)	Short description	Refs.
Process T	2,3,4,6 → 2,4,6	meta	4–34 50–60	4–34	A very restricted meta dechlorination of specific hepta- and octa-chlorobiphenyls	(Wu et al. 1997b)
Process P	2,3,4,6 → 2,3,6	Flanked para	18 30–34	18	Highly selective and removes only para chlorines flanked by at least one meta chlorine	(Quensen et al. 1990; Wu et al. 1997b)
Process N	2,3,6 → 2,6	meta	18 30–34	34	Selectively removes meta chlorines, but only those that are flanked by at least one chlorine in either the para or the ortho position	(Bedard and May 1996; Wu et al. 1997b)
Process LP	2,4,6 → 2,6 2,4 → 2	Unflanked para Unflanked para	15–30 15–30	18–22 18–22	Unflanked para dechlorination	(Wu et al. 1997b)

new concerns, because the toxicity of the transformation products may be greater than that of the parent compounds (Purkey et al. 2004). The results of a study by Buckman et al. (2006) demonstrate that fish can possibly biotransform and hydroxylate PCBs. They also suggest that it may be true that the ability of fish to biotransform PCBs is restricted, compared with birds and mammals. However, the mechanisms involved appear to be similar in all species and are dependent upon the chlorine substitution pattern of the PCBs being metabolized.

Fig. 7 summarizes the transformation processes associated with "higher chlorinated" biphenyls; potentially these are fully biodegradable, although anaerobic reductive dechlorination followed by aerobic mineralization of the lower chlorinated products is required (Field and Sierra-Alvarez 2008).

3.2 Volatilization

PCBs, especially those that have experienced significant microbial dechlorination, are very susceptible to volatilization. PCB volatilization from thin films on sand and soil was reported over 30 yr ago by Haque et al. (1974), who found minimal loss at ambient temperature. Even short drying times at room temperature resulted in significant (45–60% after 4 hr) PCB losses from sediments. Heat, air-flow (hood storage), coarse grain size, high water content, and enrichment in lower *ortho*-chlorinated congeners all were expected to increase the rate and extent of PCB volatilization (Chiarenzelli et al. 1996).

Clearly, volatilization results in differential loss of the lighter homologs, and the atmosphere is enriched by their volatilization, although subsequently they are subject to deposition from the atmosphere.

Apart from volatilization and biotransformation by microbes and higher organisms, PCBs are remarkably stable.

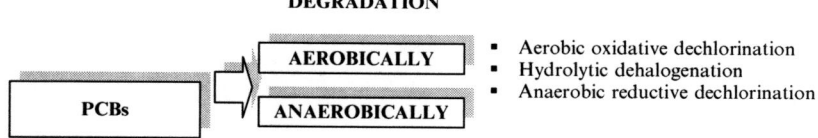

Factors that can affect PCB degradation:
- degree of chlorination – more highly chlorinated congeners are less readily degraded than less chlorinated ones
- the position of chlorine atoms on the rings
- concentration in environmental media of the congener
- type of microbial population present in media
- available nutrients, e.g., biodegradation rates decrease with high levels of organic carbon present
- pH and temperature

Fig. 7 Environmental factors that can affect the degradation of PCBs

3.3 Adsorption to Organic Matter

The environmental behavior of hydrophobic organic compounds in water is driven by partitioning between dissolved and sorbed phases. The partitioning behavior of a compound bears empirical relationships to other properties of the chemical, such as water solubility and the octanol-water partition coefficient (K_{ow}). However, actual partitioning behavior in the environment may differ significantly from such theoretical predictions (Butcher et al. 1998).

When sorbed to DOM, PCB contaminants are unavailable for uptake by organisms and, hence, become less bioavailable. In contrast, although PCBs sorbed to particulate organic matter (POM) prevent or constrain direct uptake of PCBs, these contaminants are still available to the detrital food web, which is an important pathway in rivers.

The fate of PCBs in detrital systems is dependent on the structure and origin of PCB congeners. In deeper bodies of water, fecal pellets representing POM rapidly sink to become bottom sediments (Baker et al. 1991). Cl_2 to Cl_4 PCB homologs may be present in higher concentrations in fecal material, but they are also released quickly. Baker et al. (1991) have shown that lower chlorinated PCB congeners are rapidly removed from surface waters and settle out from the water column to the sediment–water interface. At the lake floor, a large fraction of the recently settled out PCBs is released and mixed back into the water column.

Planar PCBs bind strongly to POM and are less bioavailable (Van Bavel et al. 1996). Highly chlorinated homologs sorb strongly to POM and are not assimilated easily by detritus feeders (Boese et al. 1995).

It is known that a combination of binding processes (sorption) and masstransport processes (diffusion) is responsible for the partitioning of PCBs between aqueous and solid phases and for their transport between these phases. These processes are also directly involved in, and affect the environmental fate PCBs. Precise quantitative predictions of phase speciation may allow an a priori estimate of the directly bioavailable and dissolved fractions of pollutants and their tendency for long-term dispersion in the environment. Such predictions are critical in assessing the environmental risk from PCB contamination (Gdaniec-Pietryka et al. 2007).

3.4 Bioaccumulation, Bioconcentration, and Biomagnification

Organisms can accumulate high concentrations of PCBs relative to concentrations of these substances in nonbiotic portions of the environment. This phenomenon is variously referred to as bioconcentration, bioaccumulation, and biomagnification – terms, which we try to define more precisely later (Mackay and Fraser 2000).

Some confusion exists in the literature about definitions. We adopt and follow the terms set out by Gobas and Morrison (2000).

3.4.1 Bioaccumulation

Bioaccumulation is a selective process that causes an increased chemical concentration in an organism than in the surrounding medium, and it results from uptake by all exposure routes including transport across respiratory surfaces, dermal absorption (bioconcentration), and dietary absorption (biomagnification). Bioaccumulation can thus be viewed as a combination of bioconcentration and food uptake (biomagnification).

The bioaccumulation factor (BAF) in fish is the ratio of the concentration (C_B) of the chemical in the organism to that in water, similar to that of bioconcentration factor (BCF).

$$BAF = C_B/C_{WT} \text{ or } C_B/C_{WD}$$

The most common approach for evaluating levels of bioaccumulation is to compare the levels retained by the organism, with levels in the contaminated medium in which they live (Kucklick et al. 1996).

3.4.2 Bioconcentration

Bioconcentration results from uptake of chemicals from water (usually under laboratory conditions). Uptake occurs via the respiratory surface and/or skin, and results in the chemical concentration in an organism being greater than that in the surrounding medium.

The BCF is defined as the ratio of the chemical concentration (C_B) in an organism to the total chemical concentration (C_{WT}) in the water, or to the freely dissolved chemical concentration in water C_{WD} (it only takes into account the fraction of the chemical in the water that is biologically available for uptake). The BCF is expressed as follows:

$$BCF = C_B/C_{WT} \text{ or } C_B/C_{WD}$$

Although sometimes applied to other aquatic species, the principal target organism for BCF assessment tends to be fish, primarily because of their importance as food for many species, including humans. Moreover, there are standardized testing protocols to determine BCFs for fish (European Centre for Ecotoxicology and Toxicology of Chemicals 1996). However, experimental determination of BCF values is expensive and time-consuming. Therefore, methods are used to correlate physical or structural properties of chemicals to estimate probable BCF values. For example, quantitative structure–activity relationships (QSARs) that draw on results of experimental BCF testing are used to predict bioconcentration potential of chemicals in fish (Papa et al. 2007).

3.4.3 Biomagnification

Biomagnification is the bioaccumulation of a substance up the food chain when residues are transferred from consumption of smaller organisms by larger ones in the chain. It generally refers to the sequence of processes that produces higher concentrations in organisms at higher levels in the food chain (at higher trophic levels). These processes always result in an organism having higher concentrations of a substance than is present in the organism's food. Biomagnification also results in higher concentrations of the substance than would be expected if water was the only exposure mechanism (IUPAC 1993). A biomagnification factor (BMF) can be defined as the ratio of the concentration of chemical in the organism (CB) to that in the organism's diet (CA), and it can be expressed as follows:

$$BMF = CB/CA.$$

This is the simplest definition of a BMF. It can also be described as the ratio of the observed lipid-normalized BCF to K_{OW}, which is the theoretical lipid-normalized BCF. This is equivalent to the multiplication factor above the equilibrium concentration. If this ratio is equal to or less than 1, then the compound has not been biomagnified. If the ratio exceeds 1, then the chemical is biomagnified by that factor.

The mechanism of biomagnification is not completely understood. To achieve a concentration of a chemical greater than its equilibrium value indicates that the elimination rate is slower than for chemicals that reach equilibrium. Transfer efficiencies of the chemical would affect the relative ratio of uptake and elimination. There are many factors that control the uptake and elimination of a chemical after contaminated food is consumed; these include factors specific to the chemical (solubility, K_{OW}, molecular weight and volume, and diffusion rates between organism gut, blood, and lipid pools), as well as factors specific to the organism (the feeding rate, diet preferences, assimilation rate into the gut, rate of chemical's metabolism, rate of egestion, and rate of organism growth). Because humans occupy a very high trophic level, we are particularly vulnerable to adverse health effects from exposure to chemicals that biomagnify (Bierman 1990).

A problem arises when an organism has several food sources with different concentrations. Chemicals that bioaccumulate do not necessarily biomagnify, although many papers report that PCB congeners do in fact biomagnify (Burreau et al. 2006; Nfon et al. 2008). Some early bioaccumulation models used the concept of a food-chain multiplier, which is now considered excessively simplistic (Campfens and Mackay 1997). Exposure of PCBs solely from one source only occurs in laboratory experiments (Pelka 1998). In nature, organisms are always exposed to different sources of contaminants, and therefore what happens in the field is more complex than what is reflected in laboratory studies, and it cannot easily be emulated by laboratory studies. Mass balance models are simple tools that allow evaluation of various uptake and loss processes (Mackay and Fraser 2000). A variety of mass balance models have been developed to address water quality issues in lakes, estuaries, and slow-flowing water bodies (Chapra and Reckhow 1983). The simpler models only consider advection and an overall loss due to the combined processes of volatilization, net transfer to sediment, and degradation. The rate constant for the overall loss is derived from fugacity calculations for a single segment system. The more rigorous models perform fugacity calculations for each segment and explicitly include the processes of advection, evaporation, water–sediment exchange, and degradation in both water and sediment. In this way, chemical exposure in all compartments (including equilibrium concentrations in biota) can be estimated (Warren et al. 2005, 2007). In general, these models consider the organism to be a single "box" (Fig. 8; Arnot and Gobas 2004; Mackay and Fraser 2000). These models require information about the chemicals, the organism, and associated environmental parameters (Arnot and Gobas 2004). Mackay and Fraser (2000) suggest using a tier 1 approach to screen chemicals for tendency to bioaccumulate. For those substances that are suspected of being bioaccumulative, because of their partitioning properties, a more detailed tier 2 evaluation is suggested; tier 2 uses a fish mechanistic model, which can be expressed in rate constant or fugacity formats. Such a model reveals the relative significance of gill ventilation, food uptake, egestion, and metabolism. The most detailed tier 3 evaluation is used to predict the potential for biomagnification in a food chain that involves both fish and air-breathing animals (Mackay and Fraser 2000).

PCB congeners with less *ortho*-substitution are accumulated up the food chain at a greater rate than other congeners in their homolog group (Campfens and Mackay 1997). Non-*ortho*-substituted congeners, especially those that lack adjacent unsubstituted *meta* and *para* sites and unsubstituted *ortho* and *meta* sites, are undoubtedly metabolically recalcitrant in invertebrate and vertebrate tissues (Bright et al. 1995). Changes in distributions of congeners are mainly caused by transfers among biotic compartments. There is no enrichment in higher trophic levels of mono- and non-*ortho*-substituted congeners. However, many coplanar congeners, especially very toxic PCB 77, are depleted with increasing trophic levels; PCB 77 is, therefore, almost certainly metabolized (Campfens and Mackay 1997).

Information on PCB bioaccumulation in *algae*, *invertebrates*, and fish is presented in Table 2. PCB bioaccumulation in sea gulls was described as a relatively simple two-compartment model representing partitioning between fat and blood plasma (Clark et al. 1998). But piscivorous mammals are exposed to high levels of pollutants because they feed at the top of aquatic food webs (Bremle et al. 1997).

Exposure of PCBs solely from one source only occurs in laboratory experiments. In nature, there are always multiple sources of contaminants, and therefore field results must be studied carefully. Moreover, the properties of individual PCB

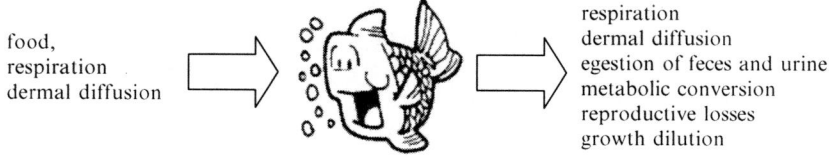

Fig. 8 Mechanisms of PCB uptake and elimination in organisms

Table 2 Factors that affect PCB bioaccumulation

Organism	Dependent factors	Lipid profile	Bioaccumulation mechanism	Uptake/sorption and desorption
Algae (*Periphyton* and *Phytoplankton*)	-Solubility	-PCBs partition to lipids in algae	-Probably a two-step mechanism, with rapid surface sorption of 40–90% within 24 hr and then a small, steady increase with transfer to interior lipids for the exposure duration (Swackhamer and Skoglund 1991)	-Desorption is significantly slower than sorption (Swackhamer and Skoglund 1991)
	-Hydrophobicity	-PCBs preferentially partition to internal neutral lipids, but those are usually a minor fraction of the total lipids, and they vary depending on growth conditions and species (Stange and Swackhamer 1994)	-Highly chlorinated congeners are associated with cell membranes (phospholipids) (Stange and Swackhamer 1994)	-Depuration from algae is very slow (Zaranko et al. 1997)
	-Molecular configuration of the congener -Growth rate[a] -Surface area and type -Content and type of lipid in the algae (Stange and Swackhamer 1994)	-Diatoms have a sticky mucus layer consisting of polysaccharides (which are polar molecules and may repel hydrophobic compounds) covering the cell surface, whereas flagellates generally lack such a layer; this may explain the significantly lower BAFs for PCBs in diatoms (Kiorboe and Hansen 1993)	-Higher BAF values for PCBs were found in the size fraction defined as phytoplankton, than in the size fraction defined as zooplankton (Magnusson et al. 2007)	-Lack enzymes for dechlorinating PCBs and no metabolism of PCBs (Hill and Napolitano 1997)
Invertebrates (*zoobenthos* and *zooplankton*)[b]	-Pore water (Forbes et al. 1998)		-Zooplankton may accumulate PCBs both across their gut epithelium, via contaminated food, and over their body surface from contact with contaminated water (Magnusson et al. 2007)	-Uptake of PCBs is rapid (Zaranko et al. 1997)
	-Physiological and ecological factors such as selective grazing and predation, different assimilation efficiencies of the contaminated food matrix, and the structure of the pelagic food web (Magnusson et al. 2007)		-The partitioning between sediment carbon and lipids in zoobenthos is referred to as the biota sediment accumulation factor (BSAF[c]) (Magnusson et al. 2006)	-Ingestion rates in deposit feeders can be greater than 100 times body w/d (Forbes et al. 1998), and they feed selectively on fine organic matter, which tends to have higher PCB contamination -Uptake of very hydrophobic compounds from sediment was observed to be one to five times greater than that predicted by equilibrium partitioning from pore water (Loonen et al. 1997) -Clearly the important pathway of exposure through the ingested detritus

(continued)

Table 2 (continued)

Organism	Dependent factors	Lipid profile	Bioaccumulation mechanism	Uptake/sorption and desorption
Fish	Fish age, size, and position in the food web (Stow and Carpenter 1994)	The lipid fraction is quite variable among species and even in the same species over the course of a year or lifetime Lipid concentrations in fish have been documented to range from 2 to 27.6% (Gerstenberger et al. 1997; Sijm and van der Linde 1995; Zaranko et al. 1997)	Bioconcentration has been modeled to mean diffusion through aqueous and lipid layer (Sijm and van der Linde 1995)	–There are two pathways for direct uptake: through the gills and through the skin –Gill exposure is a function of respiration rate, decreasing in larger fish –Dermal uptake can be significant for benthic feeders –For the species at intermediate trophic positions, the most significant loss process was loss by excretion while at higher trophic levels, growth dilution was the dominant loss process (Nfon and Cousins 2007)

[a] Phytoplankton biomass may double or triple in 1 d and periphyton turnover may be so rapid that some PCBs will not reach equilibrium, therefore, the term "bioaccumulation factor" (BAF) should be used for plants as well as animals, rather than "bioconcentration factor," which implies equilibrium (Hill and Napolitano 1997; Stange and Swackhamer 1994).

[b] Invertebrates are a critical link in both detrital and phytoplankton food webs. Higher molecular weight and more hydrophobic compounds are incorporated into sediments and are recycled by the zoobenthos (Baker et al. 1991). Lower weight, lower hydrophobicity compounds tend to be dissolved, and their uptake is enhanced in filter feeders through sorption to phytoplankton (Gilek et al. 1996).

[c] $$BSAF = \frac{tissue\ conc\ /\ tissue\ lipid}{sed\ conc\ /\ sed\ TOC}$$

where tissue conc is total PCB concentration in tissue; tissue lipid is lipid content; sed conc is total PCB concentration in sediment, and sed TOC is total organic content in sediment congeners substantially affect accumulation or degradation pathways. Empirical models only reflect one of several possible mechanisms (Antunes et al. 2007).

4 Summary

In recent decades, regulators, academia, and industry have all paid increasing attention to the crucial task of determining how xenobiotic exposures affect biota populations, communities, or entire ecosystems. For decades, PCBs have been recognized as important and potentially harmful environmental contaminants. The intrinsic properties of PCBs, such as high environmental persistence, resistance to metabolism in organisms, and tendency to accumulate in lipids have contributed to their ubiquity in environmental media and have induced concern for their toxic effects after prolonged exposure.

PCBs are bioaccumulated mainly by aquatic and terrestrial organisms and thus enter the food web. Humans and wildlife that consume contaminated organisms can also accumulate PCBs in their tissues. Such accumulation is of concern, because it may lead to body burdens of PCBs that could have adverse health effects in humans and wildlife. PCBs may affect not only individual organisms but ultimately whole ecosystems.

Moreover, PCBs are slower to biodegrade in the environment than are many other organic chemicals. The low water solubility and the low vapor pressure of PCBs, coupled with air, water, and sediment transport processes, means that they are readily transported from local or regional sites of contamination to remote areas.

PCBs are transformed mainly through microbial degradation and particularly reductive dechlorination via organisms that take them up. Metabolism by microorganisms and other animals can cause relative proportions of some congeners to increase while others decrease. Because the susceptibility of PCBs to degradation and bioaccumulation is congener-specific, the composition of PCB congener mixtures that occur in the environment differs substantially from that of the original industrial mixtures released into the environment. Generally, the less-chlorinated congeners are more water soluble, more volatile, and more likely to biodegrade. On the other hand, high-chlorinated PCBs are often more resistant to degradation and volatilization and sorb more strongly to particulate matter. Some more-chlorinated PCBs tend to bioaccumulate to greater concentrations in tissues of animals than do low-molecular-weight ones. The more-heavily chlorinated PCBs can also biomagnify in food webs. Other high-molecular-weight congeners have specific structures that render them susceptible to metabolism by such species as fish, crustacea, birds, and mammals.

In recent years, there has been substantial progress made in understanding the human health and ecological effects of PCBs and their environmental dynamics. However, risk assessments based only on the original PCB mixture that entered the environment are not sufficient to determine either (1) the persistence or toxicity of the weathered PCB mixture actually present in the environment, or (2) the risks to humans and the ecosystem posed by the weathered mixture.

In this paper, we have reviewed the status of current knowledge on PCBs with regard to environmental inputs, global distribution, and environmental fate. We conclude that to know and understand the critical environmental fate pathways for PCBs, both a combination of field studies in real ecosystems and more controlled laboratory investigations are needed. For the future, both revised and new models on how PCBs behave in the environment are needed. Finally, more information on how PCBs affect relevant physiological and behavioral characteristics of organisms that are susceptible to contamination is needed.

References

Abramowicz DA (1994) Aerobic PCB degradation and anaerobic PCB dechlorination in the environment. Res Microbiol 145:42–46.
Abramowicz DA (1995) Aerobic and anaerobic PCB biodegradation in the environment. Environ Health Perspect Suppl 103:97–100.
Alder AC, Haggblom MM, Oppenheimer SR, Young LY (1993) Reductive dechlorination of polychlorinated biphenyls in anaerobic sediments. Environ Sci Technol 27:530–538.
Antunes P, Gil O, Reis-Henriques MA (2007) Evidence for higher biomagnification factors of lower chlorinated PCBs in cultivated seabass. Sci Total Environ 377:36–44.
Arbon RE, Mincher BJ, Knighton WB (1994) Gamma-X-ray destruction of individual PCB congeners in neutral 2-propanol. Environ Sci Technol 28:2191–2196.
Arnot JA, Gobas F (2004) A food web bioaccumulation model for organic chemicals in aquatic ecosystems. Environ Toxicol Chem 23:2343–2355.
Baker JE, Eisenreich SJ, Eadie BJ (1991) Sediment trap fluxes and benthic recycling of organic carbon, polycyclic aromatic hydrocarbons, and polychlorobiphenyl congeners in Lake Superior. Environ Sci Technol 25:500–508.
Bamford HA, Poster DL, Baker JE (2000) Henry's law constants of polychlorinated biphenyl congeners and their variation with temperature. J Chem Eng Data 45:1069–1074.
Bedard DL (2003) Polychlorinated biphenyls in aquatic sediments: Environmental fate and outlook for biological treatment. In: Haggblom MM, Bossert ID (eds) Dehalogenation: Microbial processes and environmental applications. Kluwer Academic, Boston, MA, pp. 443–465.
Bedard DL, May RJ (1996) Characterization of the polychlorinated biphenyls in the sediments of Woods Pond: Evidence for microbial dechlorination of Aroclor 1260 *in situ*. Environ Sci Technol 30:237–245.
Bedard DL, Quensen III JF (1995) Microbial reductive dechlorination of polychlorinated biphenyls. In: Young LY, Cerniglia C (eds) Microbial transformation and degradation of toxic organic chemicals. Wiley-Liss Division, Wiley, New York, pp. 127–216.
Bedard DL, Van Dort HM (1997) The role of microbial PCB dechlorination in natural restoration and bioremediation. In: Sayler GS, Sanseverino J, Davis K (eds) Biotechnology in the sustainable environment. Plenum, New York, pp. 65–71.
Bedard DL, Wagner RE, Brennan MJ, Haberl ML, Brown JF Jr. (1987) Extensive degradation of Aroclors and environmentally transformed polychlorinated biphenyls by *Alcaligenes eutrophus* H850. Appl Environ Microbiol 53:1094–1102.
Bedard DL, Van Dort HM, Bunnell SC, Principe LM, DeWeerd KA, May RJ, Smullen LA (1993) Stimulation of reductive dechlorination of Aroclor 1260 contaminant in anaerobic slurries of Woods Pond sediment. In: Anaerobic dehalogenation and its environmental implications. Abstracts of the 1992 American Society for Microbiology Conference, Athens, GA. Office of Research and Development, U.S. Environmental Protection Agency, Washington, DC, pp. 19–21.
Bedard DL, Bunnell SC, Smullen LA (1996) Stimulation of microbial para-dechlorination of polychlorinated biphenyls that have persisted in Housatonic River sediment for decades. Environ Sci Technol 30:687–694.

Bedard DL, Van Dort HM, May RJ, Smullen LA (1997) Enrichment of microorganisms that sequentially meta-, para-dechlorinate the residue of Aroclor 1260 in Housatonic River sediment. Environ Sci Technol 31:3308–3313.

Benvinakatti BG, Ninnekar HZ (1992) Degradation of biphenyl by a micrococcus species. Appl Microbiol Biotechnol 38:273–275.

Berkaw M, Sowers KR, May HD (1996) Anaerobic ortho dechlorination of polychlorinated biphenyls by estuarine sediments from Baltimore Harbor. Appl Environ Microbiol 62: 2534–2539.

Bierman VJ Jr. (1990) Equilibrium partitioning and biomagnification of organic chemicals in benthic animals. Environ Sci Technol 24:1407–1412.

Boese BL, Winsor M, Lee II H, Echols S, Pelletier J, Randall R (1995) PCB congeners and hexachlorobenzene biota sediment accumulation factor for *Macoma nasuta* exposed to sediments with different total organic carbon contents. Environ Toxicol Chem 14:303–310.

Boon JP, Eijgenraam F, Everaarts JM, Duinker JC (1989) A structure–activity relationship (SAR) approach towards metabolism of PCBs in marine animals from different trophic levels. Mar Environ Res 27:159–176.

Borga K, Di Guardo A (2005) Comparing measured and predicted PCB concentrations in arctic seawater and marine biota. Sci Total Environ 342:281–300.

Boyle AW, Silvin CJ, Hassett JP, Nakas JP, Tanenbaum SW (1992) Bacterial PCB biodegradation. Biodegradation 3:285–298.

Breivik K, Sweetman A, Pacyna JM, Jones, KC (2007) Towards a global historical emission inventory for selected PCB congeners – A mass balance approach. III. An update. Sci Total Environ 377:296–307.

Bremle G, Larsson P, Helldin JO (1997) Polychlorinated biphenyls in a terrestrial predator, the Pine Marten (*Martes martes* L.). Environ Toxicol Chem 16:1779–1784.

Bright DA, Grundy SL, Reimer KJ (1995) Differential bioaccumulation of non-ortho-substituted and other PCB congeners in coastal arctic invertebrates and fish. Environ Sci Technol 29: 2504–2512.

Brown JF Jr., Wagner RE (1990) PCB movement, dechlorination, and detoxication in the Acushnet estuary. Environ Toxicol Chem 9:1215–1233.

Brown JF Jr., Wagner RE, Feng H, Bedard DL, Brennan MJ, Carnahan JC, May RJ (1987) Environmental dechlorination of PCBs. Environ Toxicol Chem 6:579–593.

Brunner S, Hornung E, Santl H, Wolff E, Piringer OG, Altschuh J, Bruggemann R (1990) Henry's law constants for polychlorinated biphenyls: Experimental determination and structure–property relationships. Environ Sci Technol 24:1751–1754.

Buckman AH, Wong CS, Chow EA, Brown SB, Solomon KR, Fisk AT (2006) Biotransformation of polychlorinated biphenyls (PCBs) and bioformation of hydroxylated PCBs in fish. Aquat Toxicol 78:176–185.

Burreau S, Zebuhr Y, Broman D, Ishaq R (2006) Biomagnification of PBDEs and PCBs in food webs from the Baltic Sea and the northern Atlantic Ocean. Sci Total Environ 366:659–672.

Butcher JB, Garvey EA, Bierman VJ Jr. (1998) Equilibrium partitioning of PCB congeners in the water column: Field measurements from the Hudson River. Chemosphere 36:3149–3166.

Campbell LM, Muir DCG, Whittle DM, Backus S, Norstrom RJ, Fisk AT (2003) Hydroxylated PCBs and other chlorinated phenolic compounds in lake trout (*Salvelinus namaycush*) blood plasma from the Great Lakes region. Environ Sci Technol 37:1720–1725.

Campfens J, Mackay D (1997) Fugacity-based model of PCB bioaccumulation in complex aquatic food webs. Environ Sci Technol 31:577–583.

Chapra SC, Reckhow KH (1983) Engineering approaches for lake management, Vol. 2: Mechanistic modeling. Butterworth/Ann Arbor Science, Woburn, MA.

Chiarenzelli J, Scrudato R, Arnold G, Wunderlich M, Rafferty D (1996) Volatilization of polychlorinated biphenyls during drying at ambient conditions. Chemosphere 33:899–911.

Chuang FW, Larson RA, Wessan N (1995) Zero-valent iron promoted dechlorination of polychlorinated biphenyls. Environ Sci Technol 29:2460–2463.

Clark T, Clark K, Paterson S, Mackay D, Norstrom RJ (1998) Wildlife monitoring, modeling, and fugacity. Environ Sci Technol 22:120–127.

U.S. Department of Health and Human Services, Public Health Service, Agency for Toxic Substances and Disease Registry (2000) Toxicological profile for polychlorinated biphenyls (PCBs). http://www.atsdr.cdc.gov/toxprofiles/tp17.pdf.

EPA (1978) U.S. Environmental Protection Agency. Support document: Draft voluntary environmental impact statement for polychlorinated biphenyls (PCBs) manufacturing, processing, distribution in commerce and use ban regulation.

Erickson MD (1997) Analytical chemistry of PCBs. Lewis Publishers/CRC Press, Boca Raton, FL/New York, pp. 37–45.

European Centre for Ecotoxicology and Toxicology of Chemicals (1996) The role of bioaccumulation in environmental risk assessment: The aquatic environment and related food webs. Technical Report 67. ECETOC, Brussels, Belgium.

Field JA, Sierra-Alvarez R (2008) Microbial transformation and degradation of polychlorinated biphenyls. Environ Pollut 155:1–12.

Fish KM, Principe JM (1994) Biotransformations of Aroclor 1242 in Hudson River test tube microcosms. Appl Environ Microbiol 60:4289–4296.

Forbes TL, Forbes VE, Giessing A, Hansen R, Kure LK (1998) Relative role of pore water versus ingested sediment in bioavailability of organic contaminants in marine sediments. Environ Toxicol Chem 17:2453–2462.

Gan DR, Berthouex PM (1994) Disappearance and crop uptake of PCBs from sludge-amended farmland. Water Environ Res 66:54–69.

Gdaniec-Pietryka M, Wolska L, Namiesnik J (2007) Physical speciation of polychlorinated biphenyls in the aquatic environment. Trends Anal Chem 26:1005–1012.

Gerstenberger SL, Gallinat MP, Dellinger JA (1997) Polychlorinated biphenyl congeners and selected organochlorines in Lake Superior fish, USA. Environ Toxicol Chem 19:2222–2228.

Gilek M, Björk M, Broman D, Kautsky N, Naf C (1996) Enhanced accumulation of PCB congeners by Baltic Sea blue mussels, *Mytilus edulis*, with increased algae enrichment. Environ Toxicol Chem 15:1597–1605.

Gobas FAPC, Morrison HA (2000) Bioconcentration and biomagnification in the aquatic environment. In: Boethling RS, Mackay D (eds) Handbook of property estimation methods for chemicals. CRC Press, Boca Raton, FL, pp. 189–231.

Gunkel G, Mast PG, Nolte C (1995) Pollution of aquatic ecosystems by polychlorinated biphenyls (PCB). Limnologica 25:321–331.

Haque R, Schmedding D, Freed V (1974) Aqueous solubility, adsorption and vapor behavior of polychlorinated biphenyl Aroclor 1254. Environ Sci Technol 8:139–142.

Hill WR, Napolitano GE (1997) PCB congener accumulation by periphyton, herbivores and omnivores. Arch Environ Contam Toxicol 32:449–455.

International Union of Pure and Applied Chemistry (IUPAC) (1993) Glossary for chemists of terms used in toxicology: Pure and applied chemistry, Vol. 65, No. 9, pp. 2003–2122 (http://sis.nlm.nih.gov/enviro/glossarymain.html. Online version posted by the U.S. National Library of Medicine).

Jafvert CT, Rogers JE (1990) Biological remediation of contaminated sediments, with special emphasis on the Great Lakes. Great Lakes National Program Office. EPA-600-991-001.

Kersting K (1975) The use of microsystems for the evaluation of the effect of toxicants. Hydrobiol Bull 9:102–108.

Kersting K (1984) Development and use of an aquatic microecosystem as a test system for toxic substances. Properties of an aquatic micro-ecosystem IV. Int Rev Ges Hydrobiol 69:567–607.

Kersting K (1991) Microecosystem state and its response to the introduction of a pesticide. Verh Int Verein Limnol 24:2309–2312.

Kersting K, Van Wijngaarden RJ (1992) Effects of chlorpyrifos on a microecosystem. Environ Toxicol Chem 11:365–372.

Kim AA, Djuraeva GT, Takhtobin KS, Kadirova M, Yadgarov HT, Zinovev PV, Abdukarimov AA (2004) Investigation of PCBs biodegradation by soil bacteria using tritium-labeled PCBs. J Radioanal Nucl Chem 259:301–304.

Kiorboe T, Hansen JLS (1993) Phytoplankton aggregate formation: Observations of patterns and mechanisms of cell sticking and the significance of exopolymeric material. J Plankton Res 15:993–1018.

Kucklick J, Harvey HR, Ostrom P, Ostrom N, Baker J (1996) Organochlorine dynamics in the pelagic food web of lake Baikal. Environ Toxicol Chem 15:1388–1400.

Li H, Drouillard KG, Bennett E, Haffner GD, Letcher RJ (2003) Plasma-associated halogenated phenolic contaminants in benthic and pelagic fish species from the Detroit River. Environ Sci Technol 37:832–839.

Lohmann R, Breivik K, Dachs J, Muir D (2007) Global fate of POPs: Current and future research directions. Environ Pollut 150:150–165.

Loonen H, Muir DCG, Parsons JR, Govers HAJ (1997) Bioaccumulation of polychlorinated dibenzo-p-dioxins in sediment by oligochaetes: Influence of exposure pathway and contact time. Environ Toxicol Chem 16:1518–1525.

Lopez MCC (2003) Determination of potentially bioaccumulating complex mixtures of organochlorine compounds in wastewater: A review. Environ Int 28:751–759.

Mackay D (1989) Modeling the long-term behavior of an organic contaminant in a large lake: Application to PCBs in Lake Ontario. J Great Lakes Res 15:283–297.

Mackay D, Fraser A (2000) Bioaccumulation of persistent organic chemicals: Mechanism and models. Environ Pollut 110:375–391.

Magnusson K, Ekelund R, Grabic R, Bergqvist PA (2006) Bioaccumulation of PCB congeners in marine benthic infauna. Mar Environ Res 21:379–395.

Magnusson K, Magnusson M, Ostberg P, Granberg M, Tiselius P (2007) Bioaccumulation of 14C-PCB 101 and 14C-PBDE 99 in the marine planktonic copepod *Calanus finmarchicus* under different food regimes. Mar Environ Res 63:67–81.

Matthews HB, Dedrick RL (1984) Pharmacokinetics of PCBs. Ann Rev Pharmacol Toxicol 24:85–103.

Nfon E, Cousins IT (2007) Modelling PCB bioaccumulation in a Baltic food web. Environ Pollut 148:73–82.

Nfon E, Cousins IT, Broman D (2008) Biomagnification of organic pollutants in benthic and pelagic marine food chains from the Baltic Sea. Sci Total Environ 397:190–204.

Papa E, Dearden JC, Gramatica P (2007) Linear QSAR regression models for the prediction of bioconcentration factors by physicochemical properties and structural theoretical molecular descriptors. Chemosphere 67:351–358.

Pelka A (1998) Bioaccumulation models and applications: Setting sediment cleanup goals in the Great Lakes. National Sediment Bioaccumulation Conference Proceedings. U.S. Environmental Protection Agency Office of Water, EPA 823-R-98-002, pp. 5-9–5-30.

Phillips KL, Sandler SI, Greene RW, Di Toro DM (2008) Quantum mechanical predictions of the Henry's law constants and their temperature dependence for the 209 polychlorinated biphenyl congeners. Environ Sci Technol 42:8412–8418.

Purkey HE, Palaninathan SK, Kent KC, Smith C, Safe SH, Sacchettini JC, Kelly JW (2004) Hydroxylated polychlorinated biphenyls selectively bind transthyretin in blood and inhibit amyloidogenesis: Rationalizing rodent PCB toxicity. Chem Biol 11:1719–1728.

Quensen III JF, Boyd SA, Tiedje JM (1990) Dechlorination of four commercial polychlorinated biphenyl mixtures (Aroclors) by anaerobic microorganisms from sediments. Appl Environ Microbiol 56:2360–2369.

Sijm DTHM, Van der Linde A (1995) Size-dependent bioconcentration kinetics of hydrophobic organic chemicals in fish based on diffusive mass transfer of allometric relationships. Environ Sci Technol 29:2769–2777.

Sokol RC, Kwon OS, Bethoney CM, Rhee GY (1994) Reductive dechlorination of polychlorinated biphenyls in St. Lawrence River sediments and variations in dechlorination characteristics. Environ Sci Technol 28:2054–2064.

Stange K, Swackhamer DL (1994) Factors affecting phytoplankton species-specific differences in accumulation of 40 polychlorinated biphenyls (PCBs). Environ Toxicol Chem 13:1849–1860.

Stow CA, Carpenter SR (1994) PCB accumulation in Lake Michigan Coho and Chinook Salmon: Individual-based models using allometric relationships. Environ Sci Technol 28:1543–1549.

Swackhamer DL, Armstrong DE (1986) Estimation of the atmospheric and nonatmospheric contributions and losses of polychlorinated biphenyls for Lake Michigan on the basis of sediment records of remote lakes. Environ Sci Technol 20:879–883.

Swackhamer DL, Skoglund RS (1991) The role of phytoplankton in the partitioning of hydrophobic organic contaminants in water. In: Baker RA (ed) Organic substances and sediments in water, Vol. 2: Processes and analytical. Lewis, Chelsea, MI, pp. 91–105.

Thomas RG (1982) Volatilization from water. In: Lyman WJ, Reehl WF, Rosenblatt DH (eds) Handbook of chemical property estimation methods. McGraw-Hill, New York, pp. 15–16.

Van Bavel B, Andersson P, Wingfors H, Ahgren J, Bergqvist PA, Norrgren L, Rappe C, Tysklind M (1996) Multivariate modeling of PCB bioaccumulation in three-spined Stickleback (*Gasterosteus aculeatus*). Environ Toxicol Chem 15:947–954.

Van Dort HM, Bedard DL (1991) Reductive ortho and meta dechlorination of a polychlorinated biphenyl congener by anaerobic microorganisms. Appl Environ Microbiol 57:1576–1578.

Warren C, Mackay D, Whelan M, Fox K (2005) Mass balance modelling of contaminants in river basins: A flexible matrix approach. Chemosphere 61:1458–1467.

Warren C, Mackay D, Whelan M, Fox K (2007) Mass balance modelling of contaminants in river basins: Application of the flexible matrix approach. Chemosphere 68:1232–1244.

White RD, Shea D, Stegeman JJ (1997) Metabolism of the aryl hydrocarbon receptor agonist 3,3,4,4-tetrachlorobiphenyl by the marine fish scup (*Stenotomus chrysops*) *in vivo* and *in vitro*. Drug Metab Dispos 25:564–572.

Wiegel J, Wu QZ (2000) Microbial reductive dehalogenation of polychlorinated biphenyls. FEMS Microbiol Ecol 32:1–15.

Wolska L, Galer K, Namiesnik J (2003) Transport and speciation of PAH's and PCB's in a river ecosystem. Pol J Environ Stud 12:105–110.

Wong CS, Lau F, Clark M, Mabury SA, Muir DCG (2002) Rainbow trout (*Oncorhynchus mykiss*) can eliminate chiral organochlorine compounds enantioselectively. Enviorn Sci Technol 36:1257–1262.

Wong CS, Mabury SA, Whittle DM, Backus SM, Teixeira C, Devault DS, Bronte CR, Muir DCG (2004) Organochlorine compounds in Lake Superior: Chiral polychlorinated biphenyls and biotransformation in the aquatic food web. Environ Sci Technol 38:84–92.

Wu Q, Bedard DL, Wiegel J (1997a) Effect of incubation temperature on the route of microbial reductive dechlorination of 2,3,4,6-tetrachlorobiphenyl in polychlorinated biphenyl (PCB)-contaminated and PCB-free freshwater sediments. Appl Environ Microbiol 63:2836–2843.

Wu Q, Bedard DL, Wiegel J (1997b) Temperature determines the pattern of anaerobic microbial dechlorination of Aroclor 1260 primed by 2,3,4,6-tetrachlorobiphenyl in Woods Pond sediment. Appl Environ Microbiol 63:4818–4825.

Zaranko DT, Griffiths RW, Kaushik NK (1997) Biomagnification of polychlorinated biphenyls through a riverine food web. Environ Toxicol Chem 16:1463–1471.

Index

1,3-Butadiene, pulmonary toxin, 52 146
2,4-Decadienal, pulmonary toxin, 52

A

Acrolein, pulmonary toxin, 50
ADHD (attention deficit hyperactivity disorder), pollutants, 2
Aerobic degradation pathway, polychlorinated biphenyls (PCBs) (diag.), 143
Aged residue effects, PCBs, 139
AIDS patients, *Pseudomonas aeruginosa*, 85
Allergens, in house dust, 18
Anaerobic degradation pathway (diag.), 143
Anesthetics, pulmonary toxins, 53
Antioxidants, benefit in pulmonary toxicity, 59
Antioxidants, pulmonary toxicity protection, 41 ff.
Anti-pollutant applications, non-thermal plasmas (NTPs) chemistry, 120
ARDS (adult respiratory distress syndrome), reactive oxygen species (ROS), 43
Asbestos, pulmonary toxin, 55
Assessment of risk, *Pseudomonas aeruginosa*, 72 ff.
Asthma reduction, home visits, 25
Asthma, allergens & microbes, 18
Asthma, induced by oxidative stress, 59

B

Babies and ADHD, pollutants, 2
Babies, house dust exposure, 1
Bacteria, free-living pseudomonads, 72
Bacteria, in house dust, 18
Bacterial infections, nosocomial pneumonia (table), 75
Benzene, pulmonary toxin, 49

Bioaccumulation factors, PCB effects (table), 151
Bioaccumulation, PCBs, 147, 148
Bio-active agents, electron transfer functionalities, 42
Bioconcentration, PCBs, 147, 148
Biodegradation, PCBs, 142
Biofilms, *Pseudomonas aeruginosa*, 92
Biomagnification, PCBs, 147, 149
Bleomycin, pulmonary toxin, 53
Burkholderia, human pathogenicity, 74
Burn-wound infections, *Pseudomonas aeruginosa*, 80

C

Cancer, PAHs (polyaromatic hydrocarbons) and house dust, 11
Carbon disulfide, pulmonary toxin, 51
Carbon monoxide, pulmonary toxin, 47
Carbon tetrachloride, pulmonary toxin, 50
Carpet contamination, lead, 4
Carpet contamination, PAHs & PCBs, 8
Carpet contamination, pesticides, 6
Carpet contamination, vacuum cleaning, 21
Carpet vs. uncovered floors, contamination and cleaning, 20
Cellular injury, ROS, 43
Chemical and biological warfare agents (CBW), disposal by NTPs chemistry, 128
Chemical contaminants, house dust (table), 9
Chemical contaminants, infant exposure, 2
Chemical mechanisms, non-thermal plasmas chemistry, 119
Children home exposure, pesticide volatility, 7
Children, blood lead levels, 6
Children, *Pseudomonas aeruginosa* infection, 83

159

Children's environmental exposure, hygiene hypothesis, 19
Chlordane, non-Hodgkins lymphoma, 8
Chlorine, pulmonary toxin, 48
Chloroform, pulmonary toxin, 50
Chronic infant health, pollutants, 2
Classification of pseudomonads, genera (table), 73
Classroom pollutants, dust, 5
Cleaning carpets vs. floors, contamination, 20
Cleaning products, safety, 23
Coal burning, PAHs in house dust, 13
Cocaine, pulmonary toxin, 56
Contaminant reduction, home visits, 25
Contaminants, in house dust, 3
Contamination by metals, classroom dust (table), 5
Contamination prevention, hand washing, 23
Contamination, vacuum cleaning, 20
COPD (chronic obstructive pulmonary disease), relative to oxidative stress, 59
Cystic fibrosis, *Pseudomonas aeruginosa*, 84

D
Diacetyl, pulmonary toxin, 52
Disease in infants, pollutants, 2
Disease transmission, *Pseudomonas aeruginosa*, 97
Disease-free water, *Pseudomonas aeruginosa (table)*, 91
Disinfection, *Pseudomonas aeruginosa*, 96
Distribution, PCBs, 137 ff.
Drinking water contamination, *Pseudomonas aeruginosa*, 87
Drinking water risks, *Pseudomonas aeruginosa* (table), 104
Dust mites, in house dust, 18

E
Ear infections, *Pseudomonas aeruginosa*, 81
EDCs (endocrine disrupting chemicals), house dust, 14, 17
Electron transfer groups, reduction potentials, 42
Electron transfer groups, relation to bio-active agents, 42
Electron transfer, pulmonary toxicity mechanism, 43
Electron transfer, pulmonary toxicity, 41 ff.
Electron transfer, redox cycling & oxidative stress, 42
Endocarditis, *Pseudomonas aeruginosa*, 77

Environmental contamination, pulmonary toxicity, 41 ff.
Environmental distribution, PCBs, 139
Environmental factors, PCB degradation effects (diag.), 146
Environmental fate of PCBs, study methods, 141
Environmental fate processes, PCBs (diag.), 142
Environmental fate, PCBs, 137 ff.
Environmental pollution, NTPs chemistry, 117 ff.
Epichlorohydrin, pulmonary toxin, 50
Ethanol, pulmonary toxin, 51
Ethylene oxide, pulmonary toxin, 46
Exposure reduction, home visits & surveys, 24
Exposure to house dust, infants, 1 ff.
Exposure to house dust, research needed, 29
Exposure, to banned pesticides, 8
Eye infections, *Pseudomonas aeruginosa*, 82

F
Farm worker exposure, pesticides, 7
Formaldehyde, pulmonary toxin, 47

G
Gaseous pollutant removal, NTPs chemistry, 121
Gastrointestinal infections, *Pseudomonas aeruginosa*, 79
Global distribution, PCBs, 137 ff., 139
Granulocytopenic patients, *Pseudomonas aeruginosa*, 86

H
Hand washing, contamination and disease prevention, 23
Health cost reduction, home visits, 25
Healthy-human infections, *Pseudomonas aeruginosa* (table), 75
Home cleaning, removing pollutants, 19
Home pollutant problems, discussion, 28
Home surveys, toxicant exposure reduction, 24
Home visits, toxicant exposure reduction, 24
Hot water pollutant extraction, carpet vacuuming, 22
House dust and allergens, infants and children, 2
House dust and exposure, research needed, 29
House dust contaminant, phthalates, 13

Index 161

House dust contaminants, pesticides & metals, 3
House dust contamination, various chemicals (table), 9, 15
House dust pollutants, exposure, 2
House dust residues, PAHs (table), 12
House dust, allergens & dust mites, 18
House dust, Bacteria, viruses, mold, 18
House dust, EDCs, 14, 17
House dust, home cleaning, 19
House dust, monitoring & sampling methods, 4
House dust, monitoring and exposure, 1 ff.
House dust, monitoring pollutants, 3
House dust, mutagens, 3
House dust, pesticides, 6
House dust, three-spot vacuum test, 21
House dust, toxicants, 6
Human health effects, *Pseudomonas aeruginosa* (table), 75
Human pathogenicity, *Pseudomonas* spp., 74
HVS3 (high-volume low-surface), pollutant sampling, 4
Hydrogen cyanide, pulmonary toxin, 47
Hydrogen sulfide, pulmonary toxin, 48
Hygiene hypothesis, children's environmental exposure, 19
Hypochlorous acid, pulmonary toxin, 48

I

Illness in spas, *Pseudomonas aeruginosa* (table), 98
Immunocompromised humans, *Pseudomonas aeruginosa* (table), 76
Indoor air pollution, house dust, 3
Indoor air purification, NTPs chemistry, 124
Infant exposure, house dust, 1 ff.
Infant hand washing, preventing contamination, 23
Infants, relative toxicant exposure, 2
Infectious diseases, hand washing, 23
Infective dose, *Pseudomonas aeruginosa*, 102
Insecticides, pulmonary toxins, 56

L

Lead in dust, child lead blood levels, 6
Lead measurement, carpets, 4
Leukemia, organochlorines in house dust, 11

M

Medical importance, *Pseudomonas* spp., 73
Meningitis, *Pseudomonas aeruginosa*, 83

Metal contamination, classroom dust (table), 5
Metal contamination, dust, 6
Metals and metal compounds, pulmonary toxins, 54
Metals, house-dust monitoring, 5
Microbial reductive dechlorination, PCBs (table), 145
Mold, in house dust, 18
Monitoring methods, house dust, 4
Monitoring pollutants, dust, 3
Mutagens, house dust, 3

N

Naphthalene, pulmonary toxin, 57
n-Hexane, pulmonary toxin, 51
Nitroaromatic compounds, pulmonary toxins, 58
Non-Hodgkins lymphoma, toxicants in carpet, 8
Nosocomial infection, *Pseudomonas aeruginosa*, 74, 76
Nosocomial pneumonia, bacterial causes (table), 75
NTPs chemistry generation, pollution abatement, 119
NTPs chemistry mechanism, pollution abatement, 119
NTPs chemistry, anti-pollutant applications, 121
NTPs chemistry, bond energies (table), 120
NTPs chemistry, disposal of CBW agents, 128
NTPs chemistry, indoor air purification, 124
NTPs chemistry, origin & definition, 118
NTPs chemistry, removing gaseous pollutants, 121
NTPs chemistry, removing odorous pollutants, 122
NTPs chemistry, removing VOC pollutants, 123
NTPs chemistry, solid waste disposal, 129
NTPs chemistry, use in sterilization, 127
NTPs chemistry, wastewater treatment, 125
NTPs, pollution abatement, 117 ff.

O

Organic matter, PCB adsorption, 147
Organochlorine contaminants, house dust (table), 9, 15
Organochlorines in house dust, leukemia, 11
Osteomyelitis, *Pseudomonas aeruginosa*, 77
Oxidative stress, asthma & illness, 59
Oxidative stress, cause for COPD

Oxidative stress, pulmonary toxicity, 42
Ozone, pulmonary toxin, 45

P
PAH (polyaromatic hydrocarbon) concentrations, house dust (table), 12, 15
PAHs, house dust, 3
PAHs and house dust, cancer, 11
PAHs in house dust, coal burning, 13
PAHs, carpet contamination, 8
PAHs, pulmonary toxins, 57
Paint thinner, pulmonary toxin, 51
Paraquat, pulmonary toxin, 56
Particulates, pulmonary toxins, 54
Pathogens, pseudomonads, 72
PCB (polychlorinated biphenyl), volatilization, environmental entry, 140
PCBs carpet contamination, 8
PCBs, adsorption to organic matter, 147
PCBs, aerobic degradation pathway (diag.), 143
PCBs, anaerobic degradation pathway (diag.), 143
PCBs, bioaccumulation, bioconcentration, biomagnification, 147
PCBs, bioaccumulation-affecting factors (table), 151
PCBs, biodegradation and transformation, 142
PCBs, determining environmental fate, 141
PCBs, distribution & environmental fate, 137 ff.
PCBs, effects of aged residues, 139
PCBs, environmental fate processes (diag.), 142
PCBs, factors affecting degradation (diag.),
PCBs, global distribution, 139
PCBs, microbial reductive dechlorination (table), 145
PCBs, occurrence and production, 138
PCBs, physicochemical properties (diag.), 139
PCBs, sources of entry (diag.), 140
PCBs, uses, 138
PCBs, volatilization, 146
Persulfate & perchlorate, pulmonary toxins, 55
Pesticide carpet contamination, track in, 7
Pesticide exposure, banned pest products, 8
Pesticide lawn use, home contamination, 7
Pesticide volatility, ingestion vs. inhalation, 7
Pesticides, farm-worker exposure, 7
Pesticides, in house dust, 6
Phenols, pulmonary toxins, 52
Phosgene, pulmonary toxin, 47
Phthalates, house dust contaminant, 13

Phthalates, pulmonary toxins, 57
Physicochemical properties, PCBs (diag.), 139
PNDEs (polybrominated diphenyl ethers), house dust contamination (table), 15
Pneumonia, *Pseudomonas aeruginosa*, 78
Pollutant abatement, NTPs chemistry mechanism, 119
Pollutant extraction, by carpet vacuuming, 22
Pollutant problems in homes, discussion, 28
Pollutants abatement, NTPs chemistry, 117 ff.
Pollutants and disease, infants, 2
Pollutants in house dust, exposure, 2
Pollutants in house dust, infants, 1 ff.
Pollutants, home cleaning, 19
Pollution abatement tool, NTPs chemistry, 118
Potable water, *Pseudomonas aeruginosa* contamination (table), 88
Pseudomonad classification, genera (table), 73
Pseudomonads, bacteria group, 72
Pseudomonas aeruginosa contamination, water, 87-91
Pseudomonas aeruginosa illness, pools & tap water, 99
Pseudomonas aeruginosa illness, spas & tubs (table), 98
Pseudomonas aeruginosa, AIDS patients, 85
Pseudomonas aeruginosa, aqueous media survival (table), 94
Pseudomonas aeruginosa, biofilms, 92
Pseudomonas aeruginosa, cancer & granulocytopenic patients, 86
Pseudomonas aeruginosa, disinfection, 96
Pseudomonas aeruginosa, drinking water risks (table), 104
Pseudomonas aeruginosa, ear & eye infections, 81
Pseudomonas aeruginosa, healthy-human infections (table), 75
Pseudomonas aeruginosa, human health effects, 75
Pseudomonas aeruginosa, immunocompromised humans (table), 76
Pseudomonas aeruginosa, in drinking water, 87
Pseudomonas aeruginosa, infection of children, 83
Pseudomonas aeruginosa, infective dose, 102
Pseudomonas aeruginosa, lung infection & cystic fibrosis, 84
Pseudomonas aeruginosa, meningitis, 83
Pseudomonas aeruginosa, nosocomial disease, 76
Pseudomonas aeruginosa, nosocomial infection, 74

Index

Pseudomonas aeruginosa, occurrence & survival, 86
Pseudomonas aeruginosa, pneumonia, 78
Pseudomonas aeruginosa, pool contamination (table), 89
Pseudomonas aeruginosa, risk assessment, 101
Pseudomonas aeruginosa, risk in water, 72 ff.
Pseudomonas aeruginosa, septicemia, endocarditis & osteomyelitis, 77
Pseudomonas aeruginosa, skin & burn-wound infections, 80
Pseudomonas aeruginosa, sources, 93
Pseudomonas aeruginosa, survival, 94
Pseudomonas aeruginosa, transmission mode, 84
Pseudomonas aeruginosa, UV disinfection (table), 96
Pseudomonas aeruginosa, water quality standards, 101
Pseudomonas aeruginosa, water transmission, 97
Pseudomonas aeruginosa-induced disease, recreational water (table), 90
Pseudomonas bacteremia, septicemia, 77
Pseudomonas spp., characteristics, 72
Pseudomonas spp., human pathogenicity, 74
Pseudomonas spp., medical importance, 73
Pulmonary injury, ROS, 43
Pulmonary toxin, ozone, 45
Pulmonary toxins, ethylene oxide, sarin & sulfur mustard, 46
Pulmonary toxins, formaldehyde, phosgene, carbon monoxide & hydrogen cyanide, 47
Pulmonary toxins, hydrogen sulfide, sulfur dioxide, chlorine & hypochlorous acid, 48
Pulmonary toxins, metals & particulates, 54
Pulmonary toxins, structures (diag.), 44
Pulmonary toxicity mechanism, electron transfer, 43
Pulmonary toxicity, antioxidant protection, 41 ff.
Pulmonary toxicity, electron transfer, 41 ff.
Pulmonary toxicity, ROS and oxidative stress, 42
Pulmonary toxins, acrolein, epichlorohydrin, chloroform & carbon tetrachloride, 50
Pulmonary toxins, anesthetics & therapeutic agents, 53
Pulmonary toxins, asbestos, silica, persulfate & perchlorate, 55
Pulmonary toxins, benzene, toluene, styrene, trichloroethylene, 49
Pulmonary toxins, diacetyl, phenols, 1,3-butadiene, 2,4-decadienal, 52
Pulmonary toxins, ethanol, n-hexane, paint thinner, carbon disulfide, 51
Pulmonary toxins, naphthalene, PAHs, phthalates & quinones, 57
Pulmonary toxins, nitroaromatic compounds & radiation, 58
Pulmonary toxins, pesticides, tobacco, cocaine, 56

Q
Quinones, pulmonary toxins, 57

R
Radiation, pulmonary toxin, 58
Radicals and pulmonary toxicity, possible mechanism, 41 ff.
Reactive nitrogen species, similarity to ROS, 45
Reactive oxygen species (ROS), cellular injury & ARDS, 43
ROS, pulmonary toxicity, 42
Recreational water, *Pseudomonas aeruginosa* contamination & disease, 88, 90
Redox cycling and electron transfer, toxicity mechanism, 43
Reducing infant exposure, house dust, 1 ff.
Risk assessment, *Pseudomonas aeruginosa*, 101
Risk assessment, *Pseudomonas aeruginosa*, 72 ff.

S
Safer cleaning products, home cleaning, 23
Samplers, high volume for pesticides, 4
Sampling and monitoring, house dust, 4
Septicemia, *Pseudomonas* bacteremia, 77
Silica, pulmonary toxin, 55
Skin infections, *Pseudomonas aeruginosa*, 80
Soil degradation, PCBs (diag.), 146
Solid waste disposal, NTPs chemistry, 129
Sources of entry, PCBs (diag.), 140
Sterilization, using NTPs chemistry, 127
Styrene, pulmonary toxin, 49
Sulfur dioxide, pulmonary toxin, 48
Surface water, *Pseudomonas aeruginosa* contamination, 89

T
Transformation, PCBs, 142
Therapeutic agents, pulmonary toxins, 53
Three-spot test, vacuuming dust contaminants, 21
Tobacco, pulmonary toxin, 56
Toluene, pulmonary toxin, 49

Toxicant exposure in infants, relative intensity, 2
Toxicants in carpet, non-Hodgkins lymphoma, 8
Toxicants, metals in dust, 6
Toxicity mechanism, electron transfer-reactive oxygen species-oxidative stress, 42
Trichloroethylene, pulmonary toxin, 49
Urinary tract infections, *Pseudomonas aeruginosa*, 78

U
UV light, *Pseudomonas aeruginosa* disinfection (table) 96

V
Vacuum cleaners and contaminants, cleaning, 20
Viruses, in house dust, 18
VOC pollutant removal, NTPs chemistry, 123
Volatilization of PCBs, environmental entry, 140
Volatilization, PCBs, 146

W
Wastewater treatment, NTPs chemistry, 125
Water quality standards, *Pseudomonas aeruginosa*, 101

Breinigsville, PA USA
26 August 2009
222997BV00006B/11/P